Beginning Digital Electronics Through Projects

Beginning Digital Electronics Through Projects

Andrew Singmin

Newnes

Boston Oxford Auckland Johannesburg Melbourne New Delhi

Newnes is an imprint of Butterworth–Heinemann.

Copyright © 2001 by Butterworth–Heinemann

 A member of the Reed Elsevier group

All rights reserved.

No part of this publication may be reproduced, stored in a retrieval system, or transmitted in any form or by any means, electronic, mechanical, photocopying, recording, or otherwise, without the prior written permission of the publisher.

Recognizing the importance of preserving what has been written, Butterworth–Heinemann prints its books on acid-free paper whenever possible.

Butterworth–Heinemann supports the efforts of American Forests and the Global ReLeaf program in its campaign for the betterment of trees, forests, and our environment.

Library of Congress Cataloging-in-Publication Data
Singmin, Andrew, 1945–
    Beginning digital electronics through projects / Andrew Singmin.
      p. cm.
    Includes index.
    ISBN 0-7506-7269-2 (pbk.: alk. paper)
    1. Digital electronics—Amateurs' manuals. I. Title.
TK9965.S5443 2000
621.381—dc21   00-064639

British Library Cataloguing-in-Publication Data
A catalogue record for this book is available from the British Library.
The publisher offers special discounts on bulk orders of this book.

For information, please contact:

Manager of Special Sales
Butterworth-Heinemann
225 Wildwood Avenue
Woburn, MA 01801-2041
Tel: 781-904-2500
Fax: 781-904-2620

For information on all Newnes publications available, contact our World Wide Web home page at: http://www.newnespress.com

10 9 8 7 6 5 4 3 2 1

Printed in the United States of America

For a very special Suzie V—*thank you for the line dance lessons.*

Gloucestershire County Library	
991998528 7	
Askews	23·3-01
621.381	£14.99

Contents

Preface xi

1 Introduction 1

2 Fundamentals of Analog Circuits and Introduction to Digital Circuits 7

Fundamentals of Analog Circuits: From the Crystal Set to the Integrated Circuit 8
Digital Electronics 17

3 Testing Principles and Test Equipment 27

Test Principles 27
Test Equipment 31
Project 1: LM 555 Demo Oscillator 41

4 Digital Circuit Truth Tables 45

AND Gate 45
OR Gate 48
NAND (NOT-AND) Gate 50
NOR (NOT-OR) Gate 53
EX OR (Exclusive OR) Gate 55
EX NOR (Exclusive NOR) Gate 56

5 Components Review and Projects 2–8 59

Electronic Components 59
The Basic Power Supply 60
Resistors 60
Capacitors 66

Switches 69
Jack Plugs and Sockets 70
Project Start 70
Project 2: LF TTL Pulse Generator 71
Project 3: High-Low State Display 77
Project 4: Pulse Speed Reducer 81
Project 5: Threshold Level Detector 84
Project 6: Hi-Lo Trip Detector 88
Project 7: LED Pulse Stretcher 90
Project 8: dc Level Shifter 93

6 Schmitt Trigger Circuits: Projects 9–13 97

Introduction 97
Project 9: Schmitt Pulse Oscillator 97
Project 10: Turn-On Delay Schmitt 100
Project 11: Turn-Off Delay Schmitt 103
Project 12: Schmitt Triangle Generator 105
Project 13: Schmitt Switch Debouncer 108

7 Versatile Digital ICs: Projects 14–20 111

Project 14: NOR Gate Demo 111
Project 15: SCR Latch Demo 114
Project 16: NOR Gate Latch 117
Project 17: NOR Gate Metronome 120
Project 18: 74LS122 Monostable 123
Project 19: 74LS75 Quad Latch 126
Project 20: CD4072 Quad Switcher 129

8 Digital Support Circuits: Projects 21–26 133

Project 21: TTL Relay Driver 133
Project 22: High-Power Transistor Driver 137
Project 23: Power Output FET 141
Project 24: TTL Driver Buffer 143
Project 25: Passive Five-Step Voltage Indicator 146
Project 26: Active Five-Step Voltage Indicator 148

9 Special Power-Supply Circuits: Projects 27–38 153

Project 27: RF Oscillator Stable 6.2-Volt Supply 153
Project 28: Stable Five-Volt Source 157
Project 29: Battery or Adapter Supply 159
Project 30: Diode Bridge Circuit Protector 161

Project 31: Regulator Protector   163
Project 32: Reversible Voltage Source   164
Project 33: Jack Socket Tutorial   166
Project 34: Meter Overload Protector   169
Project 35: Jack Socket-Power Switch   171
Project 36: Zener Five-Volt Supply   173
Project 37: RC Integrator Differentiator   174
Project 38: Variable Resistor Substitute   177

10 Single-Chip IC FM Receiver: Project 39   181

Tuned Radio-Frequency (TRF) Receiver   182
Superheterodyne Receiver   182
Circuit Description   183
Construction Tips   184
Test Setup   187
Parts List   187

11 FM Low-Power Transmitter: Project 40   189

Introduction   189
Circuit Description   190
Stabilizing RF Oscillator Drift   193
Construction Tips   194
Test Setup   195
Parts List   197

Index   199

Preface

Putting *Beginning Digital Electronics Through Projects* together began with ideas about a follow-up sequel to my first book, *Beginning Electronics Through Projects* (subsequently retitled for the second edition *Beginning Analog Electronics Through Projects*). There is a definite continuity between the *Analog* and *Digital* volumes that I wanted to maintain. With that aim in mind, the structure for this *Digital* project evolved.

If I were to choose one outstanding difference between *Analog* and *Digital* that I had personally found, it would undoubtedly be that among the different analog integrated circuits available, you can still differentiate and make sense out of their prime functions—an operational amplifier IC, a power amplifier IC, a voltage regulator IC. These are fairly self-evident explanations. But when you consider digital circuits, such as an OR gate IC, a Schmitt trigger IC, or a latch IC, these functions don't spring easily to mind. From the immense list of standard digital ICs, it is difficult therefore to narrow down to a few devices that can be used as an experimental setup from which to start acquiring a digital knowledge base. The analog situation is easy; the LM 741 operational amplifier, LM 555 timer, and LM 386 audio power IC serve (as seen in my earlier books, *Beginning Electronics Through Projects* and *Practical Audio Amplifier Circuit Projects*) as the basis for a wealth of introductory circuits, which are also fairly well known to most electronics enthusiasts. There is no digital equivalent! So I decided that the main theme of this book should be a good baseline selection of commonly available digital ICs.

At the end of reading this book, you will have become acquainted with a number of good (i.e., useful) digital devices. There is no special reason attached to my selection of these particular digital devices, other than their being ones I have worked with. My approach puts these devices in a practical-build situation. All the circuits are real, proven circuits, built from scratch and tested before reaching the versions you see before you now.

If my circuits are "real," what then are "unreal" circuits? So-called "application notes" are what I place in that category. Manufacturers' data books are used by the electronics design community, particularly in reference to the operating, performance, and capability specifications of devices (integrated circuits). You can find numerous application notes for digital circuits in any of these data books. At first glance these appear to be especially nice circuits to build, but in reality they are only suggested circuits put together by applications engineers, showcasing the manufacturer's devices. Rarely have they been actually built. Ask anyone attempting to get the suggested application-note circuit to work correctly, and you'll most likely get the answer that there are a lot more components needed before the circuit can be considered to be a real, working one. I've discovered this the hard way, believing that I've just found a terrific new circuit idea. As a result of the frustration of working with application notes, I tend to avoid them when looking for real circuit ideas. Specifically, operational amplifier application notes are shown operating from a dual power supply; if you want to run it from a single nine-volt supply, more important components are needed.

Even more critically, if you're running the basic configuration in either a noninverting or inverting mode, the components will differ (see my *Practical Audio Amplifier Circuit Projects* for such examples). Digital circuit application notes, on the other hand, don't show pull-up or pull-down resistors—you'll see what I mean later in this book. The circuits that you see in my books are all built from the ground up. I have tested them thoroughly before committing them to paper and found them to be 100 percent functional. You'll see that in a lot of cases I've included in the circuit schematics a jack socket at the input and output (as appropriate). It doesn't affect the circuit operation, obviously, but (to me) it does emphasize that the circuit is a practical one.

A stand-alone digital circuit really doesn't do too much, as contrasted to my favorite analog example, an audio amplifier. It's clear what an amplifier does—it makes audio sounds (music) louder. A typical digital circuit involves a logic state change—so what? You can't empathize with a logic state change. There is a certain degree of abstraction, therefore, associated with digital circuits, and it makes understanding them a little more difficult for the beginner. This book is designed to make your introduction to digital circuits easier.

In spite of all that I've said so far, there is one digital-associated feature that comes out on top over analog. With any basic digital circuits, such as gates, we generally want to monitor the output change of state in response to an input change of state. The key word here is "monitor." As this is simply a dc voltage binary condition, that is, it's either high or low, we can employ the simplest monitor around—a plain LED (with current limiter resistor)—to perform that function. There's no simple equivalent for the analog case. So the advantages are shared out. The state changes are latched, not transients, making the monitor requirements so easy.

This was really a fun book to put together, from the very early concept stage, through development, and finally seeing the copy version make its appearance. There's a terrific satisfaction in seeing to fruition what had started out as a concept proposal. I particularly enjoyed having the freedom to come up with a theme and structure for this book that reflected the way I wanted to achieve my goal. When you've reached the end of this book, I hope you will have advanced your expertise in digital electronics and had fun in building up and experimenting with these circuits. I'd like to thank my friends Candy Hall and Pam Chester at Newnes for their closeness during the writing of this book.

Andrew Singmin
April 2000

CHAPTER **1**

Introduction

My first publication, *Beginning Electronics Through Projects*, dealt with analog circuits; as more appropriate to the contents therein, the second edition was retitled *Beginning Analog Electronics Through Projects*. That first publication gently led you through an introduction to electronics by building simple projects, starting with the "very easy" and gradually developing into more complex circuits. The circuits were generally analog in nature, that is, mainly amplifier and oscillator circuits. Here in the follow-up volume (*Beginning Digital Electronics Through Projects*), we introduce you to a new series of circuits. This book is mainly digital in content but includes also two innovative RF (radio frequency) circuits, a FM (frequency-modulated) single IC (integrated circuit) receiver, and a transistor-based FM transistor transmitter. Information on high current drivers and latches expands your working-knowledge base for applying the digital circuitry to your own enhanced building blocks. A selection of circuit configurations using a digital monostable, latch, and OR gate will let you see how actual devices can be intercoupled.

Analog circuits, of the type described in volume 1 (*Beginning Electronics Through Projects*), are simpler to work with and perform functions, such as amplification, that can be very easily verified; for example, you know an amplifier is correctly functioning when there's sound coming out of the speakers. It is not so, however, with digital circuits, where voltage levels are basically just switching between two levels, high and low. From that description, you would be inclined to wonder where the interest is in digital circuits. Well, by taking a controlled run through digital circuits, starting in chapter 1 with the basics, you'll see how digital circuitry progresses. As you become familiar with them, first through theory and then through practical circuit builds, the value of digital circuits should become clearer. All microprocessor circuits found in computer systems, however complex they are, are based on basic digital circuits. The knowledge gained here also has value later on, when you start finding out about the challenges of more advanced, more complex digital circuitry. A more basic treatment of electronics can be found in my previous publication, *Beginning Electronics Through Projects*.

All the circuit descriptions provided here have, in keeping with my philosophy, been built and tested to confirm that they function correctly. All necessary components and correct interconnection locations for unused IC pins are included in these original circuit designs. The chapter contents are broken down as follows:

Chapter 2. Fundamentals of Analog Circuits and Introduction to Digital Circuits

In this opening chapter, we start with a brief review of the fundamentals of analog circuits. The fundamental concepts of analog circuits are relatively easy to understand, and simple amplifier projects are the best types to start with. Why? Because everyone is familiar with radios, cassette players, CD players, and TVs. What is the one feature common to all of these electronic products? They produce sound, which we use our ears to process. That is, we can all tell if the sound is loud or soft, high or low, fast or slow (if it's music). It is very easy, therefore, to verify that an amplifier, which is a sound processor and one of the most common examples of an analog electronics circuit, is working or not.

Let's say you have a sound source, such as an electric guitar, for example. Unplugged (i.e., when it's not coupled to an amplifier), the sound is just what you hear coming from the strings. It's intelligible but soft (of low volume). Now plug it into an amplifier, and you can make the sound as loud as you want it just by adjusting the volume control. Therefore, you can easily tell what the amplifier does—it makes the sound louder, that is, amplifies it. There's not much difficulty, even for the nonelectronics person, in appreciating what an amplifier does.

Digital circuits are much more difficult to comprehend, and there's no easy circuit equivalent to the analog amplifier. But we've got to make a start somewhere, and that's where chapter 2 leads off. Once the details of basic analog circuit concepts are covered, the chapter starts with digital logic levels, binary numbers, and digital electronic technology. Because there are different types of digital logic families in use, it is important to know what the fundamental differences are; this topic closes the chapter. The practical and fun part of working with simple digital circuits comes later, but for now some technical background is necessary.

Chapter 3. Testing Principles and Test Equipment

As soon as you've built a project, what is the first thing you have to do? Test it! Without testing it, you'll never know if it's working or not—so your understanding of how you go about testing and how you use test equipment is important. This chapter therefore makes a start with the principles of testing.

Introduction **3**

Any test strategy is designed with one purpose in mind—to find and rectify fault symptoms. Test equipment varies from the simple multimeter to the more complex oscilloscope. Each piece of test equipment has its unique purpose. Fortunately, there's a lot you can do with the basic multimeter, but to progress to more thorough levels of signal measurement, the oscilloscope is going to be needed in a lot of cases. You will also need a signal source, as a stimulus for interrogating a number of different circuit types. That's what the function generator does; it produces a number of different signal source types. After the basic coverage of test principles and test equipment, there's a simple demonstration circuit build. Nothing beats having an actual circuit to test, and that's the purpose here. This simple digital oscillator circuit provides you with a wealth of signal points to measure and at the same time have fun with the actual circuit build.

Chapter 4. Digital Circuit Truth Tables

Digital circuits always have what are called "truth tables." Digital circuits operate on the principle that the output sees the results of logic level changes. Logic level changes would occur at the input of a digital circuit, and the effect is monitored at the output. A truth table is just a convenient tabular form for representing the various input state changes and the resultant effects on the output. Each type of basic logic circuit, called a gate, has its own truth table. This chapter covers the most common types of digital gates, their truth tables, and a breakdown of what it all means. In order to understand realistic chains of logic circuits (which is what digital circuits are generally composed of), it is necessary to understand what the most fundamental level of digital circuit (the gate) does (how it operates); that is the purpose of this chapter.

Chapter 5. Components Review and Projects 2–8

If you've been anxiously waiting to get building, this chapter is where it starts. For beginners new to project builds or even those wanting a refresher, the section begins with a review of components. You're going to be handling components in any build, be it analog or digital—the components are the same, so this is a good starting place. The circuits that follow are all very basic, easy to build, but very versatile in use. In my view, circuit usefulness does not have to be exclusively a function of circuit complexity; in other words, less is more.

In all there are seven very useful projects described here, are all of them nice examples of digital circuitry.

Chapter 6. Schmitt Trigger Circuits: Projects 9–13

One of the most distinctive digital circuits, out of the many available, is the Schmitt trigger. Its uniqueness lies in the fact that the same basic circuit can be used to provide a vast range of very different digital circuit functions.

4 BEGINNING DIGITAL ELECTRONICS THROUGH PROJECTS

The most notable example, and the one that is most often encountered, is the use of the Schmitt trigger to clean up noisy pulses. Digital circuits are renowned for requiring clean pulses to work with. This is not an unusual requirement, since digital circuits are basically based on gates, and gates operate on the principle of specific output states defined against specific input states. Change the input state, and the output state changes. Noisy signals have unpredictable voltage levels. If these appear at the input to a logic gate, the output becomes unpredictable—and that can't be tolerated, since by the very definition of digital circuits, we expect to have predictable states! Mechanical switches are a quick and easy means to apply varying input logic levels, but they also generate unwanted and unacceptable transients. The Schmitt trigger, as you'll see in this chapter, cleans up those transients. Additionally, and among other uses, the Schmitt is a versatile pulse generator source. The build circuits are:

Schmitt pulse oscillator
Turn-on delay Schmitt
Turn-off delay Schmitt
Schmitt triangle generator
Schmitt switch debouncer

Chapter 7. Versatile Digital ICs: Projects 14–20

Because of the profusion of digital ICs available, wondering where to start can be quite a headache. Not only that, once you've made a choice, digital ICs don't actually do very much as stand-alone circuits—you've got to configure them into a useful circuit. The advantage of this chapter, therefore, is it offers you a selection of what I consider to be easily available, simple-to-use digital ICs, and also circuit schematics for illustrating how these digital ICs can be put to work. The circuit complexity builds up considerably as you progress into the chapter, but each circuit shown utilizes just one IC. Digital ICs often contain more than one gate—four gates are not uncommon—hence the reason for the circuit complexity. The build circuits are:

NOR gate demo
SCR latch demo
NOR gate latch
NOR gate metronome
74LS122 monostable
74LS75 quad latch
CD4072 quad switcher

Chapter 8. Digital Support Circuits: Projects 21–26

Real circuits often require more circuitry than textbooks show to get them going. This chapter looks at a variety of very simple yet effective circuits that I call "supporting circuits." A simple digital circuit, for example, looks good on paper and even on the test bench, but when it comes to a real application, where it's needed to feed into a real load—it won't work. Why? The drive capability of a regular TTL gate is very low, so it can't be used directly to drive any load of significance. You're going to need some sort of driver circuit. That's what this chapter is all about; also, there are two unusual circuits at the end. The build circuits are:

TTL relay driver
High power transistor driver
Power output FET
TTL driver buffer
Passive five-step indicator
Active five-step indicator

Chapter 9. Special Power Supply Circuits: Projects 27–38

All circuits, analog or digital, require a power supply to operate from. Usually, the simplest way of supplying the power is a nine-volt battery. But there are instances where a little more is required. One topic always leads onto another, so there's also a nice section on jack sockets. Why? Jack sockets are the final connection to the outside world. As usual, there are a few surprise circuits thrown in for good measure at the end. The build circuits are:

RF oscillator stable supply
Stable five-volt source
Battery or adapter supply
Diode bridge protector
Regulator protector
Reversible voltage source
Jack socket tutorial
Meter overload protector
Zener five-volt supply
Jack socket power switch
RC integrator differentiator
Variable resistor substitute

Chapter 10. Single-Chip IC FM Receiver: Project 39

We close with two special circuit builds that are RF (radio frequency) based. RF circuits are notorious for being difficult to build, set up, and make function. This and the next RF circuit show how a practical circuit can be affected. Anyone can put together a RF circuit of sorts, but making it work is another matter. That's (as you've already seen) the focus of this book (and all my other ones): taking the practical approach to circuit builds and turning them into real, actual, working projects! This is a real, working FM receiver build, based on the Signetics TDA7000 IC.

Chapter 11. FM Low-Power Transmitter: Project 40

The second RF circuit is based on a single transistor as the oscillator source. You often see this basic circuit in electronics magazines. Some variants are somewhat difficult, as I've found in the past, to get working. What sets this version apart is the details I've provided on finding the transmitter frequency. This is, I've found, the most tricky part of getting the circuit going—and I've never seen anyone else highlight this. The usual trimmer capacitor used for frequency tuning requires you to use considerable care in locating your frequency. It's not as simple as a volume control potentiometer, which you just turn up to get a louder sound output. Once you've got the circuit going, it's very well behaved.

CHAPTER **2**

Fundamentals of Analog Circuits and Introduction to Digital Circuits

Analog electronics technology introduced in my first book, *Beginning Electronics Through Projects*, is the simplest type to get started with. Almost invariably, a beginner's introduction to electronics will involve analog electronics. Analog electronics is very tangible, in that you can see or hear the electronics "working." Take, for example, an audio amplifier; the sound can be heard from the speaker, and that's a clear indication, if that's the project you've just finished, of what the circuit does. Even a nonelectronics person is able to appreciate that.

Analog signals that make up this branch of electronics can vary continuously in magnitude, going from a zero value to a maximum value that is limited only by the system in use—so it can be seen as having no limit. If you were to measure analog signals, let's say with a voltmeter, you would find an infinite excursion of values. Most analog signals are dynamic—that is, they constantly and randomly vary. Where they are static, they're referred to as dc signals, but they're still analog signals. A nine-volt battery could be thought of as an example of a static analog signal. It's static because it's not varying over time, or at least varying so slowly that to us (humans) it seems to be staying still. But we know that batteries eventually die, even those on the shelf; if we were to compress the time scale, we'd see that the monitored battery voltage would be varying (even to us, now) and hence analog in nature. Any signal that's analog in nature is also referred to as an ac signal. Circuits that are analog in nature can generally be recognized by the fact that there are coupling capacitors at the input and the output of, say, amplifiers.

What else is analog in nature? Oscillators are an example of a case where although the frequency is constant, the amplitudinal variation over time is changing in a cyclic nature—that is to say, it is continually undulating, going from a zero amplitude to a maximum, then dropping past zero to a negative minimum, and repeating the cycle forever as it moves through zero again. (This is the pattern for a sine wave.) Of course, you can also have an oscillator that's a square wave (it still produces an audible tone); the wave form is not a

7

continuum this time but goes through a stepwise excursion from a zero voltage baseline level to a maximum level, returning to zero and repeating forever. This is an interesting point, as we see that oscillators can be both analog and digital in origin.

There is however, another interesting and subtle observation about the analog and digital oscillator. Circuits centered around oscillators are ac coupled out—that is, we're only interested in the ac component. By passing a signal that is effectively digital in origin through a capacitor, we convert it to an analog signal. So although the square-wave oscillator is digital, we're only interested in its analog property.

Generally when we're considering digital electronics we want to retain the dc component information in slow-moving signals, and therefore digital circuits are devoid of the usual coupling capacitors seen in ac circuits. Capacitors block dc signals.

Fundamentals of Analog Circuits: From the Crystal Set to the Integrated Circuit

The first appearance of electronics is often represented as the crystal set—simplicity in itself, with the minimum of parts, yet capable of apparently magical powers, drawing in mysterious signals from the sky. With no external supply of power needed, the crystal set is indeed the only example of getting something for nothing. Who in their youth has not been awed by the mystique of sitting up in the twilight hours, headphones tightly in place, searching across the medium-wave broadcast bands, parallel metal blades meshing within a sturdy capacitor framework? Seeing those blades mesh is so much more rewarding than modern-day faceless tuning, with an all-singing, all-dancing remote. Four components, that's all you needed, and with four components, what construction issues were there? No, even a simple twisting together of connections would produce awe-inspiring signals. No switch was needed, because of the lack of a power source—amazing!

From simple designs to sophisticated variations, crystal sets even today retain their fascination, and first-timers use them to cut their teeth on the exciting prospects of electronics. But with the dawning the vacuum-tube age, power was to be had. A single warm, orange-glowing tube, standing proud, drove your headphones to untold heights of radio-signal enjoyment. There's something comforting about a glowing tube that no solid state counterpart can ever equal. Obsolete? Far from it—serious, dedicated audiophiles and musicians hold tube amplifiers in a state of almost unwarranted reverence. Germanium transistors at their first appearance couldn't hold a candle to tube performance. A one-tube amplifier provided more than satisfactory sound reproduction, whereas a one-transistor design couldn't be realized.

The transistor designs that did materialize had benefits of low-battery-power operation, something at which tubes were severely disadvantaged.

Attempts at some form of radio portability with tube designs were really dragged down by the huge physical battery power (as high as 90 volts!) needed for drivers. So transistor radios, powered by the ever-popular nine-volt battery, took over the crown, and the mass appeal of the tube started to spiral downwards.

The replacement of germanium by silicon didn't draw as much attention as the first integrated circuit's emergence. Transistor-based circuitry until then was certainly capable of delivering circuit solutions within the grasp of the amateur constructor, but high component counts and specialized component requirements made it a tedious, albeit enjoyable, task. The integrated circuit changed all that, bringing with it circuit capabilities previously unheard of that could be built with amazing ease. The performance capability of the LM 386 audio power amplifier, with its low current consumption—it's driven by one nine-volt battery—and with only two external components, is awe-inspiring.

What you've seen so far is the compressed emergence of electronics milestones, from the crystal set to the early integrated circuits. Most electronics applications, especially those encountered by the beginner and hobbyist, were analog in nature. Analog signals, typically represented by amplifiers, worked with real signals, signals varying over time, signals representative of how they actually originated in life. The simple carbon microphone converts speech (changes in sound pressure) to electrical signals; the signals are processed through an amplifier chain and emerge later, larger in amplitude than when they started out, but still the same thing. That's an analog signal processor. Analog signals can take on any amplitude from zero to infinity; there are no unallowed values, the signals span a continuum spread of values.

In simple terms, analog electronics can be defined as covering signals that are analog in nature *and* retain their analog nature after processing. The reason for the emphasis is that analog-origin signals can be converted into digital (the subject of this book), and by my definition the technology is then analog no more.

Do all signals originate as analog? Is there also a digital universe out there? Let's look at typical signals that are familiar. Signals come from transducers, which are just devices that convert energy of one form into another. Here are some transducers:

1. Microphones converting sound pressure waves into electrical signals. The actual process conversion is through the perturbation of carbon granules, which means a change in resistance. This dynamic resistance is used as the driving force for signal generation. As we know from a simple potentiometer circuit arrangement, we can cause the output voltage (appearing across the wiper terminal) to vary in sync with a variation in resistance caused by varying the

potentiometer shaft. So it's not difficult to conceive of converting a changing resistance to a changing signal source. Some other microphones, like the crystal microphone, actually develop a voltage as opposed to a resistance change.
2. A thermistor is a temperature sensing device, with its resistance typically decreasing as the temperature increases. Temperature can, obviously, be detected by merely measuring resistance. The thermocouple also is used as a temperature sensor, but here there is an actual voltage generated.
3. The accelerometer is an acceleration (velocity change) sensing device, a physical mechanism that generates a changing resistance in response to acceleration changes. The resistance changes are translated, as usual, into a signal.
4. The photoconductive cell converts light intensity into changing electrical resistance, and the same conversion applies as before.

So far, the well known transducers quoted are all analog in nature—that is, the signals are continuous. Perhaps there are no digital signals out there?

So, all electronics that maintain the analog nature of an originating signal (analog) can be described as analog electronics. Analog is by far simpler than digital to visualize, and it's much more tangible also—in the sense that our human senses are much more attuned to analog signals. Look at the analog and digital multimeter as an example, and you'll see what I mean. Consider a changing waveform, audio type, like the output from a music source. When it is appropriately fed to an analog meter to display its changing dynamics, the visualization is obvious. Our eyes can spot highs, lows, transients, and flats very easily—but switch over to the digital display, and our eyes can't empathize the same way. A visual display of changing numbers is cold. I guess humans are analog in nature.

What you read about in my first book, on introductory electronics, is all analog in nature. They show circuits at the very beginning level, covering mainly practical, real, working amplifiers for driving common audio signals. There's also a tone generator included, a little bit of an anomaly in itself, since the output is analog—it's a continuously amplitude varying signal, but it's digital at the signal origin.

The confusion can be easily explained by the following breakdown. Let's take first a classic analog sine wave source that's used as the basis for an audio signal tone generator. Assume we've got a middle-of-the-road audio frequency of 1 kHz. This is a good mid-range frequency, given that the audio spectrum spans approximately 15 Hz to 20 kHz. The 1 kHz sine wave would come from regular signal generator. Now slowly decrease the frequency. The signal tone drops. Go on until it starts to get really "bassy"; we're down now to frequencies in the sub-100 Hz region. As you go lower to the limit of the response of your audio system, the signal tone will die away smoothly. Now repeat the scenario with the square-wave generator (this is like our tone generator). As the frequency drops, there will come a point when you start hearing a distinctive on-off click. This is the sound of the square wave (which, incidentally, is a digital signal) pulsing on and off. Because the human ear does not resolve frequencies above a certain minimum, a square wave above a certain frequency sounds like a continuum.

The acid test to distinguish between an analog and a digital signal is to drop the frequency. If it's solid, it's analog. If it drops, it's digital. When the frequency of a true analog tone signal is reduced, the tone starts to reduce, ultimately tailing off to a deep low-frequency rumble. When the frequency of digital tone signal is reduced, the tone first starts to reduce, but at some stage, when the frequency is really low (a few Hz) you'll see (if you've got this on an oscilloscope display) the waveform switching or jumping back and forth between a high and low value—unlike the true analog case, where signal always remains a true continuum, no matter how low the frequency is reduced. That's the key difference between an analog and a digital signal. An analog system therefore can either use an analog or digital signal and, through the processing stages, result in an analog output.

To make an even finer distinction, the actual frequency of operation also affects our definition of the system. Say we have a digital square-wave signal at 1 kHz (a nice audio tone, albeit somewhat raspy sounding) that's fed through an audio amplifier and into a speaker. Nothing unusual—that's an analog processing system. But now say the signal is not 1 kHz any more but really low, say, 1 Hz. That's not an audio frequency, and the amplifier of before can't handle it; neither can the speaker. So the output device would have to be different and appropriate. It could simply be an LED; this is now a digital system. We haven't changed the input signal stage, just dropped the frequency drastically.

Having almost exhausted the options on this issue, let's now stand back and see what it is we (I) define as analog electronics. It's a system (a string of electronic components performing a particular function) that, when all is considered, results in an analog output. Whew! We finally got there.

Looking at my first book, we can see that all of its topics fit under the category of analog electronics. We have (in my book) the audio pre-amplifier, audio power amplifier, tone generator, guitar amplifier, and microphone pre-amplifier. Not surprisingly, the simplest category of analog electronics is audio devices (amplifiers of one sort or another). The output from such analog circuits is so easily verified that it is the ideal type of circuit for beginners to get familiar with. Ohm's Law, the backbone of all electronics current, voltage, resistance calculations, is also so much more at home with analog electronics than with digital.

Take the simplest case, a powered-up LED (light-emitting diode) with its current-limiting resistor. Say we start with a convenient nine-volt power supply and a 4.7 kohm limiting resistor. Ohms law says that if we decrease the resistance, the current will rise. So drop the 4.7 kohm to 1 kohm; the LED is brighter because of the higher current flow. That's a nice, simple example of analog electronics. If you were to replace the fixed resistor by a variable resistor and had a current meter in series to monitor the current, you could see the current changing smoothly as you rotated the potentiometer shaft. There are no quirky digital jumps to contend with (but more of that later, of course, as this is a book on digital circuits).

But for the time being, under the review of analog electronics, we can explore some more of what we already know of analog electronics. Take the operational amplifier circuit, the simplest type being the inverting mode type. Armed with an oscilloscope for signal-display purposes, and using, as we've learned earlier, either an audio square or sine-wave signal source, we can see that the basic analog amplifier setup must have input and output capacitors before and after the amplifier, to block out any dc component. That's another way in which analog's simpler than digital; the dc always goes in analog, and we don't ever have to worry it, it's value (amplitude) or polarity (positive or negative). Once more, analog's simplicity emphasizes its attractiveness. Beginner's guides or starter texts on electronics always begin with analog, never digital. It is essential to have a good grounding in analog before embarking onto the newer and very different digital world.

Usually, and especially at the introductory level, we are mainly interested in amplifying signals, first low-level signals into higher-level signals, and then power amplifying these to some useful work. Thereafter, the most popular and common treatments are signal generation and signal filtering. So let's look at these in a little more detail.

Signal Pre-amplification

Signal pre-amplification, when assisted by the ever-popular operational amplifier, such as the LM 741, is without a doubt the first and most popular encounter in the analog electronics world. Why? Well, electronics is very much based on ultimately producing a useful output, such as a TV signal you can see, or a radio signal you can listen to. Having stated that fact, we can next go on to state that if you're going to produce these useful signals, there's got to be a source somewhere, and there is. At this stage its makeup doesn't matter much, just that it's there. What else do we know, or can we guess, about signal sources? They're always weak. Isn't that a strange thought? If such signals were huge in strength, we wouldn't need so much amplification, and then perhaps there wouldn't be the need to develop the knowledge base we have. But that's not the case. The modern integrated circuit allows us to explore ways and means of getting amplification easily, unlike the days of old when transistor-based amplifiers, with their temperamental characteristics, had to be built. The IC is stable—remarkably so! You choose what are essentially two noncritical components, and you've got amplification. It is a far cry from the intricacies and unknowns of trying to put together a transistor-based equivalent. Does anyone even do it this way now, I wonder? It seems unlikely. Why go to mono when there's stereo, and why go to tape when there's CD?

You can do wonders with operational amplifier–based circuits as pre-amplifiers. Although signal amplitudes can be greatly and easily amplified, there's more required before they can be put to good use. Don't forget, we always want to have a useful function at the end of our electronics journey (remember the TV signal and the radio). Life being what it is, though, there's nothing simple.

Signal Power Amplification

Signal power amplification now starts to make its appearance. It still amplifies, like its pre-amplifier counterpart, but the power tag gives us a clue. Like the power of a 427 muscle V-8, power is the ability to drive, and in the case of electronics, this is the power to drive loads. Loads require more power to drive when they're low impedance, and—you guessed it—real-life loads are low impedance. We never seem to get away from trials thrown up by nature. A low-impedance load takes power, or current, to perform. This current has to come from somewhere, and that somewhere is the power supply. The plain, simple rule is that the more power you want, the more the source has to be able to supply. Power is product of voltage and current, and nothing comes for free. The more volume you want from an audio system, the more current is drawn. At some point the battery is unable to supply your needs, and then it's time to draw power from the 110 V AC line socket. It's not taken directly, of course; through what is called a power supply circuit, raw alternating 60 Hz is converted into, say, 9 V DC.

The basic conversion runs like this. The line 110 V AC is reduced down, using a transformer, to a value close to what the required final dc value is. The reduced 110 V AC is then rectified, through a process called either full-wave or half-wave rectification. What this does is separate the ac sine wave's positive and negative peaks into the positive peaks. With a half-wave rectification system, we get the positive peak, a gap, then the positive peak, a gap, a positive peak, and so on. When the system is a full-wave rectifier, the gaps are filled with positive peaks, so we get a smoother waveform, as the gaps are missing. These peaks are not much good as they stand; fed into a very large capacitor, the peaks start to approach a dc value with a little ripple (that is, the small portion of 60 Hz AC left). More down-stage filtering takes care of the remaining ripple until we get a clean, hum-free dc.

Another, less involved, way to get more power is to use a line adapter. Typically, to reproduce a nine-volt battery, something like a 12 V adapter is used. The stated 12 V is only unloaded—that is when you couple in a load, the voltage drops to some value determined by the load. That's not really satisfactory. A much better option is to use an integrated circuit regulator, which will provide a nine-volt output regardless of the load conditions. This is mentioned by way of example only, as nine-volt regulators are less common than five-volt types. The reason for this is that digital circuits run off five volts, and the industry caters to what is most prevalent.

However, another point to consider with using adapters is that the hum level is very high, unacceptably high for audio use. That means the 60 Hz ripple has to be reduced. Inside a typical line adapter, there's very little in the way of capacitive filtering. Large capacitors to filter 60 Hz cost money, and that translates to less profit for the manufacturer, so out goes the capacitor. Most intended applications, in any case, are not audio; they're mainly calculators. A large external capacitor—several 1000 µF capacitors, rated at 25 V each, wired in parallel—will do the trick. If you're running audio with an adapter, just keep adding 1000 µF parallel capacitors and you'll hear the hum disappearing. Why? The capacitors have an impedance that falls as capacitance increases, as the capacitor is a shunt path.

Signal Generation

Signal generation has a lot of uses, most specifically for measurements and fault tracing. Ideally, the sine-wave shape is the best to use, but practicalities dictate that we can't have everything. The square wave turns out to be the more often utilized signal source. The four (my listing only) best integrated circuits around with regards to their ultimate usefulness are: the LM 741 generic plain vanilla operational amplifier, the LM 386 audio power amp on a chip, the LM 555 the ultimate timer, and the LM 324 quad op-amp pack. The LM 555 timer chip is designed to run in two modes, single shot and free running. "Free running" is another way of saying oscillating—but it only oscillates in square-wave mode. There are, incidentally, sine wave chip oscillators, but unless they are as freely available as the LM 555 and equally simple to configure, I can't put them in the same league of usefulness. So there is only one signal generator chip contender, the LM 555.

That's why the square wave turns out to be more used than the sine wave. It's simply a case of availability. But with a little circuit cunning (using just three passive components), we can get a pretty good approximation to the sine wave and so lose some of the raspiness of the square wave. This is the triangular wave. As far as audio is concerned, it's not bad, and a lot better than the square in any case. A square wave has a lot of harmonics. Harmonics are additional frequencies that are higher-order multiples of the fundamental, or main, frequency. Say you had 1 kHz as the main frequency. The second harmonic would be two times 1 kHz, which is 2 kHz; the third harmonics is three times 1 kHz, which is 3 kHz; and so on. A square wave can be described mathematically by an equation that shows the wave is composed directly of odd-order (three, five, seven, etc.) harmonics. If you played with a fundamental sine wave, added a third harmonic and replotted that, then added the fifth harmonic and added that too, you'd see the original sine wave starting to look square.

As a means of measurement making, signal sources are fed to the input (if it's an amplifier we're investigating), and an oscilloscope is hooked up to measure the relative amplitudes between the output and the input. That ratio's the gain. If a 10-millivolt signal is fed to an amplifier and it produces a 1-volt signal, we've got a gain of ×100 at that frequency. Our measurement is only as good as what we start with. We need to have a clean signal (sine wave or square wave); the response of the unit under test (the amplifier) to other than the two reference signals is unknown. We also need to have control over the amplitude (to take into account the differing input-level requirements) and over the frequency (usually the audio band covers most requirements). A less critical requirement for signals arises when they're used as tracers for fault tracking. In some applications, particularly amplifier chains, injecting a signal at the input and tracing back in stages will identify where the breakdown occurs. By and large, any signal frequency in the audio range and having a few 10 s of millivolts amplitude will serve perfectly for this application.

Signal Filtering

Signal filtering is a little more subtle, sometimes a mite tougher to spot. Invariably though, especially in very simple setups, the property that capacitive impedance decreases with increasing frequency is used to full effect in controlling upper-frequency bounds. Signal filtering requires a little knowledge of what existed prior to any applied filtering. Under a nonfiltered environment, we expect the gain to be constant over the frequency band of interest. In the simplest case, we can assume that this is the audio frequency band; typically this spans 15 Hz to 20 kHz. Therefore, if the gain is, say, ×10 at a convenient reference point (I generally define this to be 1 kHz), we expect that at the top (20 kHz) and low (15 Hz) ends, the gain will be constant at ×10. In reality, there are usually some circuit restrictions limiting this range, and it's more likely to be somewhere between a few 100 Hz and, say, a few kHz above 10 kHz. The actual range we define is not important. Signal filtering will modify the frequency response in a number of ways, but the most common are high-pass filtering and low-pass filtering; one is just the complement of the other. For this explanation, we'll assume it's a low-pass filter we're dealing with.

An oddity with the definition of filters is worth explaining here. The low-pass filter, as its name implies, passes low frequencies; so far so good. If that's the case, and since, as we know, filters alter the frequency response of a circuit, then by default, a low-pass filter has to have some effect on the high frequencies. Sounds a little odd, but is that the case? Yes. A low-pass filter could in my opinion be more accurately defined as a high-cut filter. But filter terminology has stuck with the word "pass," and hence we always have this terminology anomaly. After a while you get used to what is really meant by a low-pass filter.

A low-pass filter is the simplest filter type to analyze, especially with the version that you'll see used here. Some textbooks on filters tend to complicate the treatment so much, what with equations and graphs, that they make understanding the basic operation very difficult. So, with a low-pass filter setup, we know that at a certain midpoint reference frequency (generally you can use 1 kHz for that frequency without too much problem), the gain is constant. What the actual value is not of importance. As the frequency increases, we'll find at some point (that is, frequency), the gain starts to drop, and for the very simplest filter case, the gain will fall off at a constant rate with increasing frequency. That means that knowing the slope of the fall-off, you can calculate the gain at any frequency.

Finally, we'll come to what the low-pass filter actually looks like. First, remember what the standard inverting operational amplifier circuit looks like; it's the one with the feedback resistor and input resistor for setting the gain. The gain, as we know, is just the ratio of the feedback to the input resistor. For a 100 kohm feedback and 10 kohm input, the gain is simply 100 kohm/10 kohm = 10. Imagine now that we place across the feedback resistor a small capacitor, something like 100 pF. The feedback total resistance is now the parallel combination of the feedback resistor and feedback shunt capacitor. As the capacitor's reactance (ac resistance) varies with frequency, the parallel combination is more accurately called an impedance. The units are still the same, though (in ohms). As the frequency increases, the capacitive reactance drops. The total parallel impedance also drops (the total impedance of a parallel circuit is always less than the smallest individual resistive/reactive component). That, then, means that the gain falls with increasing frequency—and that is where we started with the definition of a low-pass filter concept. We can see therefore that adding a simple capacitor across the feedback resistor gives us a basic low-pass filter circuit.

It's not so basic, though, that it is not practical—far from it. This happens to be my favorite filter circuit (low-pass) for explaining filters to beginners. It is really simple, yet effective, and it is hardly seen in filter texts, which in the main seem to follow the philosophy that more complex is better. I totally disagree! If you can't understand the basic principles because the treatment is too difficult, what's the use? You cannot get simpler than a one-component (additional) filter! Incidentally, the circuits in my books always use an easily obtainable single-voltage supply, a nine-volt battery. You will also find other treatments of filter circuits made more difficult by the use of a dual voltage supply. The components, especially around the biasing network, are different for the dual and single voltage cases. From a practical point of view, the single-voltage design is more useful for a simple beginner's use, because at that level there is little to be gained by having dual supplies. Again, you will find few texts (other than mine) using a single power supply for driving operational amplifiers.

If you want to read more about the analog concepts and work through some practical builds, check out my previous volumes, *Beginning Electronics Though Projects* and *Practical Audio Amplifier Circuit Projects*.

Digital Electronics

We've seen now four common examples of basic analog circuit functions. A lot of complex circuits can be broken down into these four basic building blocks, simplifying the understanding process. The analog process has been around for a long while and will continue to be there in spite of the omnipresent digital world. Because of the prevalence of computer technology, digital technology (the basis of computer technology) probably has been brought to the forefront. Digital electronics works on a basis of two discrete values, called logic states, most commonly represented by two specific voltage levels. Generally a high logic level is associated with a high voltage level, and a low logic level with a low voltage level. The logic high state is also represented by the unit "1" and the logic low state by the unit "0." By generating the appropriate sequence of logic bits we can thus make up logic values. As there are only two logic states and hence two voltage levels, this system is known as a binary system. That makes a nice introduction into our new section on digital electronics.

Digital Logic Levels

All digital electronics is founded on a system of "1"s and "0"s or high and low voltages to formulate a series of words that ultimately represent an analog equivalent. Known formally as a binary coded decimal or BCD system, this system represents decimal numbers in a binary format.

Table 2-1 Four-Bit BCD Coding System

Decimal Digit	Binary Code			
	A	B	C	D
0	0	0	0	0
1	0	0	0	1
2	0	0	1	0
3	0	0	1	1
4	0	1	0	0
5	0	1	0	1
6	0	1	1	0
7	0	1	1	1
8	1	0	0	0
9	1	0	0	1
10	1	0	1	0
11	1	0	1	1
12	1	1	0	0
13	1	1	0	1
14	1	1	1	0
15	1	1	1	1

To see more closely the generation pattern for converting decimal numbers to binary, take a look at a three-bit system below. The number of decimal numbers represented is seen when all the bit positions are filled in with "1"s, beginning with 000.

Table 2-2 3-Bit BCD Coding System

Decimal Digit	Binary Code		
	A	B	C
0	0	0	0
1	0	0	1
2	0	1	0
3	0	1	1
4	1	0	0
5	1	0	1
6	1	1	0
7	1	1	1

Rather than printing out long lists every time, the number of decimals represented by any n-bit system is simply given by:

$$\text{Maximum decimal number} = 2^n$$

where n is the number of bits in the system being considered, and 2 is the base of the number system in use. The base (also called the radix) is the total number of distinct symbols used in a system. In the usual decimal system, the radix is ten, made up of the symbols 0, 1, 2, 3, 4, 5, 6, 7, 8, 9. These ten symbols enable any number to be generated on the convenient base-10 system. In the binary numbering system the radix is 2, made up of the symbols 0, 1. As before, any number can be generated just using these two symbols (0,1). Another way of looking at the definition of base, or radix, is that it is the total number of characters made available to each position of a numbering system. For the decimal system, these positions would correspond either to the units column, the tens column, or the hundreds column, and so on. For the binary system, the same positions would represent the units column, or the twos column, or the fours column, or the eights column, and so on.

The four bits are read starting from the farthest right and then incrementing bit by bit to the left, i.e., D to C to B to A. Each bit's letter designation has a certain value: $D = 1, C = 2, B = 4, A = 8$. These values originate from the use of the 2-bit binary code. So $2^0 = 1, 2^1 = 2, 2^2 = 2 \times 2 = 4$, and $2^3 = 2 \times 2 \times 2 = 8$.

The binary number system has just two values, 1 and 0, and uses 2 as its base. Our more familiar decimal system uses the numbers 0, 1, 2, 3, 4, 5, 6, 7, 8, and 9, and it has a base of 10; that is, once the number 10 has been reached, the system of counting starts again. The number 123 in the decimal system says that we have (going from right to left) 3 single units, plus 2 units of 10s (making 20), plus 1 unit of hundreds (making 100), the total being 3 + 20 + 100, which is one hundred twenty-three, or 123. No individual number can be greater than 9, after which it rolls forward to 0. Because of our familiarity with the decimal system, we don't even think of what the numbers mean when we see "123" written.

But try this with a binary system: What does 0111 mean in a four-bit binary-coded decimal system? Not so obvious now! We know that each bit cannot be greater than 1. That explains why you see only the numbers 0 and 1 in a BCD system. Having been given 0111, we need to decipher each bit from right to left, just as we did for the 123 decimal example. Take the first bit farthest to the right; that bit's a "1," so that equals 1. The next bit going to the left is "1," so that's equal to 2. Next to that position the "1" equals 4. The last bit is "0," so that's equal to 0. Adding these all up we have: 1 + 2 + 4 + 0 = 7—which is what we have in Table 2-1 above. Here's a second example. Take the BCD word 1110. What is its decimal equivalent value? Working from right to left, the equivalent decimals are 0 + 2 + 4 + 8 = 14.

Remember in the binary system, the units increment from right to left, as 1, 2, 4, 8. For a four-bit binary coded system, the more bits we have, the bigger the number that can be represented.

For a two-bit system, the maximum number represented is
$2 \times 2 = 4$ numbers (0 to 3)
For a three-bit system, the maximum number represented is
$2 \times 2 \times 2 = 8$ numbers (0 to 7)
For a four-bit system, the maximum number represented is
$2 \times 2 \times 2 \times 2 = 16$ numbers (0 to 15)
For a five-bit system, the maximum number represented is
$2 \times 2 \times 2 \times 2 \times 2 = 32$ numbers (0 to 31)
For a six-bit system, the maximum number represented is
$2 \times 2 \times 2 \times 2 \times 2 \times 2 = 64$ numbers (0 to 63)
For a seven-bit system, the maximum number represented is
$2 \times 2 \times 2 \times 2 \times 2 \times 2 \times 2 = 128$ numbers (0 to 127)
For an eight-bit system, the maximum number represented is
$2 \times 2 \times 2 \times 2 \times 2 \times 2 \times 2 \times 2 = 256$ numbers (0 to 255).

Digital Electronics Technology

Digital logic technology can be broadly split into two groups: transistor-transistor logic (TTL) and complementary metal-oxide semiconductor (CMOS). A further split is often seen, with metal-gate CMOS and high-speed CMOS—the characteristics of which are seen in Table 2-3 below. TTL is the more commonly known logic technology, operating off a +5 V supply voltage. TTL is robust (not static sensitive) and based on bipolar (transistor) technology. Enhancements to first-generation TTL reduced power and increased speed of operation, enabled by the use of Schottky diodes in the design, such that standard TTL was now supplemented with higher-performance LSTTL (low-power Schottky TTL). In fact, LSTTL has generally superseded standard TTL. LSTTL technology has considerable output drive capability.

CMOS technology, in contrast, uses field effect transistors as opposed to the bipolar transistors in TTL (an important difference), and both the design structure and method of operation of CMOS are totally different from TTL. The advantage of using CMOS over TTL is its very low power dissipation (essential for battery-powered miniaturized circuits). As improvements in CMOS technology have accumulated, the key focus has been in the reduction of the supply voltage from five volts to three volts.

Although high speed might always be a desire, it brings an assortment of attendant issues: increased noise generation, higher power consumed, and increased component count and complexity.

TTL requires a supply voltage that is restricted to a very narrow range, 4.5 V to 5.5 V, in order for the device's performance to remain within the manufacturer's specification. CMOS is much more tolerant; for a TTL-equivalent high-speed variant, the range is less restrictive, from two to six volts. The low operating voltage results in a reduced power consumption; operating speed, however, is reduced.

The drive capability, i.e., the ability to drive a number of loads without degradation in speed, is highest with standard LSTTL, lowest in metal-gate CMOS, and somewhere in between with high-speed CMOS.

TTL consumes moderate power when operating, the consumption remaining constant over frequency (to about 10 MHz) and then rapidly escalating as the frequency rises. TTL power is characteristically in the milliwatt region. CMOS power consumption, in contrast, rapidly increases almost linearly with frequency but drops to zero in the zero frequency (quiescent) mode. CMOS power is characteristically in the microwatt region. Although CMOS power consumption can rise to milliwatts at the maximum operating frequency, the average power consumption is very small, since the percentage of the total number of devices on at any time is small. Power consumption can also be reduced by reducing the supply voltage, as power is proportional to the square of the supply voltage. TTL devices are more efficient than CMOS at higher frequencies.

Unused inputs for LSTTL should be connected to Vcc (5 V) and not left floating.

CMOS

CMOS for one manufacturer started out as the well known MC 14000 metal-gate series, which later succumbed to the increasing demand for higher speed. Microprocessors, especially, required very-high-speed devices, and consequently the technology of choice was the faster LSTTL device, which had a speed gain of around eight to one. CMOS's benefits of low power, attractive for portable equipment, had unfortunately to be discarded. Inevitably, however, the efforts in CMOS research focused on a technology with both low power and high speed. The HSCMOS series (high-speed CMOS) utilizes silicon-gate technology and yields a device size that is one-half that of the metal-gate technology. The smaller device results in savings in silicon. High-speed CMOS is compatible with LSTTL and metal-gate CMOS, allowing immediate performance upgrades to be achieved in many cases by plug-in replacements.

CMOS devices are susceptible to damage from electrostatic discharge, but with the inclusion of protection circuits at the input, static discharges in excess of 2000 V can be tolerated. Devices damaged by static can exhibit total failure, i.e., a low impedance path between Vcc and ground, and consequently be unlikely to respond to a regular signal stimulus at the input terminal. Intermittent failures or diminished performance can be traced to less severe static charge levels. Leakage currents are also likely to increase as a result of static damage.

For the maximum protection against damage, it is advisable to take particular care when handling CMOS devices.

1. A proper work surface should be used with a conductive surface connected to ground.
2. Wrist straps properly connected to ground must be worn before handling any CMOS device.
3. All unused device inputs must be connected to ground.
4. Store and transport CMOS devices in antistatic or conductive containers.

The two most common high-speed CMOS categories are:

HC series: This is a high-speed CMOS with CMOS input switching levels and buffered outputs. HC devices can interface directly into LSTTL.

HCT series: This is a high-speed CMOS with LSTTL to CMOS input and output interfacing capability. LSTTL can interface directly into HCT.

Both metal-gate CMOS and high-speed CMOS types are pinout compatible with LSTTL (low power Schottky TTL) devices.

Table 2-3 LSTTL versus CMOS Performances

Parameter	LSTTL	CMOS (metal-gate, very high speed)	
Voltage supply	4.5 to 5.5 V	3 to 18 V	2 to 6 V
Power/gate	5 mW	0.0006 mW	0.003 mW
Speed (toggle)	33 MHz	4 MHz	45 MHz
Output drive current	8 mA	1 mA	4 mA

22 BEGINNING DIGITAL ELECTRONICS THROUGH PROJECTS

The CMOS metal-gate series is known as the 14000 series by one manufacturer, and the high-speed CMOS series by the prefix HC or HCU. The critical differences are shown in Table 2-3, where it is clearly seen that metal-gate has a very wide supply-operating range of 3 to 18 V, compared to high-speed CMOS's two to six volts. Power consumption is less by a factor of five in favor of metal-gate CMOS, but its output drive loses out to high-speed CMOS by a factor of four. Speed is naturally in favor of high-speed CMOS, by a factor of over ten. Comparisons with LSTTL clearly show the significantly lower power consumption of CMOS in comparison to TTL. Clearly, there is a compromise (when is there ever not?) to be considered when choosing technologies.

TTL, although somewhat eclipsed by the lower power of CMOS, is still probably the most popular form of digital logic. TTL logic is easy to use, robust (no static precautions needed), low cost, has medium- to high-speed operation, and has a good output drive capability. TTL power consumption is reduced further through the introduction of the Schottky series (LSTTL), which provides an improvement by a factor of five between LSTTL and TTL. The Schottky diode clamping arrangement prevents the device from going into saturation.

LSTTL Switching Levels

With a supply voltage of 5 volts and the input voltage at zero, the output sits at a high state (4.5 V). As the input voltage is increased, the output will drop to a low state (0.2 V) as the input crosses 0.8 V. This input voltage (0.8 V) sets the point below which the output will not switch over from a high to low state. It is defined as V_{IL}, i.e., the maximum low-level input voltage, or the worst-case voltage that is recognized by the device as a low-input state. Similarly, of course, we would expect there to be an input voltage that sets the point above which the output will not switch over from a low to high state. This is defined as V_{IH}, i.e, the minimum high-level input voltage, or the worst-case voltage that is recognized by the device as a high-input state. V_{IH} is two volts for LSTTL.

Typical Digital Devices

Introductory analog electronics can be confined to just a few humble integrated circuits that are sufficient to develop a comfortable grasp of the fundamentals of circuit schematics and project constructions. Analog integrated circuits like the LM 741 operational amplifier, LM 555 timer circuit, and LM 386 audio power amplifier are all that are needed to build an entire range of very useful devices. There is, of course, a vast portfolio of other analog devices, as evidenced by a quick glance at any analog IC manufacturer's product line. But the point here is that we don't need to go beyond these three basic ICs. If you were to take a look at just the operational amplifier section of a manufacturer's offerings, you'd see listing for high-speed op amps, low-noise op amps, and specialist op amps with each and every function optimized. Generally, of course, one function would be optimized at the expense of another, but that's an inevitable fact of life; performance in one area rarely comes without compromise in another area. Regardless, these are still op amps, and the basic inverting mode amplifier still operates on the fundamental need for a feedback resistor and an input resistor to set the gain. Once you've learned the fundamental amplifier principles with the workhorse LM 741, you're set, and there's no need to learn anything different with other op amps. A low-noise op amp, for example, could be a direct plug-in replacement for the LM 741. Design a basic pre-amplifier using the LM 741, say, with quite a high gain of ×100, and you'll probably notice a fair amount of noise when the complete audio chain (power amplifier, speaker) is finished. The LM 741 is not designed for serious hi-fi audio quality, but a low-noise equivalent IC that is audio specific can be replaced for the socketed LM 741 (a good reason to use IC sockets); the improvement in noise reduction is remarkable.

Digital integrated circuits, in contrast, are much more numerous, even at the basic level of understanding. That is because there are so many different digital IC designs, each catering to a specific logic design. Digital circuit design takes the form of a number of complex, interrelated logic gates. It is not possible to characterize a digital circuit as easily as we would an analog circuit. Analog circuits can be described nicely—pre-amplifiers, power amplifiers, sine-wave oscillators, square-wave oscillators, microphone amplifiers, guitar amplifiers, high-pass filters, low-pass filters, and so on. Each description will tell any electronics designer what we mean. Other than the individual digital logic blocks (the ICs) themselves, which could be so described, the final digital circuit design has no such simple tag. That difference makes the whole issue of digital electronics so much harder to comprehend.

The way to start the understanding process is to describe a number of basic digital building-block ICs, that is, fundamental digital ICs that are most commonly found in circuit designs. We can only describe the ICs themselves, as the circuits have no easy descriptive tag. For the beginner to digital electronics, the consequence of such a difference (from analog electronics) is that circuits of a digital nature might seem less attractive. For that reason it is strongly recommended that any starter electronics be made with the analog rather than digital form. It is, of course, assumed that an enthusiast buying this book will already have a basic analog grounding, not so much as a necessity to understanding digital, but because learning electronics will be far more rewarding with an analog start, given the superior intrinsic attractiveness of analog circuitry.

But to return to digital circuits—because digital circuit design is based on the use and interaction of logic gates, we're going to start with the most common types of logic gates found in digital electronics. These form the backbone of all designs. The simplest starter gate is the AND gate. As the name (AND) might imply, we are going to consider two entities, something AND something; as might be expected, we get a logic result based on the interaction of these two somethings.

We might have something like this: entity A AND entity B when combined give rise to some new entity C. Sounds a little abstract, but it'll be clearer as we progress. In any case, the whole of digital electronics technology is founded on gates, so we need to give this our attention. We have introduced two entities A and B that, when combined, gives rise to a new entity C. So it is not unlikely, from an electronics viewpoint, that this scenario should be represented by some as-yet unnamed device having two inputs, which we'll label input A and input B, not forgetting the output, which we'll label output C. So far, so good.

But there is another very important aspect to characterizing this device block. We're dealing with digital circuits; recall that a digital circuit can take on only two discrete values; either it's at a logic low or logic high. So now, armed with this information, it's appropriate to infer that input A can take on two logic states, either a logic low or a logic high. If input A can assume these states, then so can input B, and extending the argument, output C also. In fact, any terminal in a digital circuit can take on a logic low or logic high state. Within this one paragraph we've come a long way in setting up the basis of digital electronics. All digital electronics operate on the principle of terminal nodes taking on certain logic states (defined by the designer) and the effect of these logic states propagating throughout the circuit. Large circuits become difficult to analyze, but with the simpler examples described in this book, we can make a good start in understanding the logic flow process.

Logic Family Developments

As with any technology, we can expect advancements to be made over time as processes improve. Design requirements also drive device enhancements. High speed and low power consumption are the ideal characteristics. Here's how TTL evolved.

- 74 Series: TTL. This is the first series of logic IC introduced; all subsequent logic enhancements are based upon it. Speed and power among other characteristics, are collectively optimized for most applications. The greatest variety of logic devices is available in this logic family. Power dissipation is typically about 10 mW per gate.
- 74L: low-power TTL. This low-power family of logic devices has basically the same characteristics as the standard TTL 74 series, except power dissipation is reduced dramatically to less than a tenth of that of the standard TTL. As with everything else, low power is achieved at the expense of device speed (which is reduced by about a factor of three with respect to the standard TTL). Where low power is needed, the 1 mW per gate dissipation makes it the ideal choice.
- 74LS: low-power Schottky. Low-power Schottky has the same speed as TTL using a special process incorporating a Schottky diode (hence the Schottky in the title), but with a significant power reduction, a fifth of TTL.

74S: Schottky. Schottky is faster than TTL with a speed gain of 3 × TTL, but at the expense of a power consumption that is twice that of TTL.

74ALS: advanced low-power Schottky. Advanced low-power Schottky has a speed gain of 2.5 over TTL, and a low power consumption, a tenth that of TTL.

74AS: advanced Schottky. This has the fastest speed, six times TTL, with a moderate increase in power consumption, 1.5 that of TTL.

Table 2-4 TTL Series Progressions

	74 Series: TTL	74L Series: Low Power TTL	74LS Series: Low Power TTL	74S Series: Schottky	74ALS Series: Advanced Low Power Schottky	74AS Series: Advanced Schottky
Speed	TTL	1/3 × TTL	TTL	3 × TTL	2.5 × TTL	6 × TTL
Power	TTL	1/10 × TTL	1/5 × TTL	2 × TTL	1/10 × TTL	1.5 × TTL

Frequently Used 74 Series TTL Logic Devices

When faced with a proliferation of digital devices, it's difficult to know where to start. Unlike the analog favorites, LM 741, etc., digital selection is a little tougher because of the choices to make. So as a help, from the huge array of digital logic devices available, my favorites, the most useful, are listed below. These are the ones that would be most likely to show up in circuit designs. If you want to experiment with digital logic devices, this could be a nice list to choose from. The power requirement is five volts. The ideal, and also the simplest, solution if you're running from a nine-volt battery is to use a five-volt regulator IC that takes in nine volts and gives out a stable, load-independent five-volt output.

- 7400: Quad 2-input NAND gate
- 7402: Quad 2-input NOR gate
- 7404: Hex inverter
- 7408: Quad 2-input AND gate
- 7410: Triple 3-input NAND gate
- 7413: Dual 4-input Schmitt trigger
- 7414: Hex Schmitt inverter
- 7432: Quad 2-input OR gate
- 7442: BCD to decimal decoder
- 7446: BCD to seven segment decoder
- 7468: Dual decade counter
- 7473: Dual J-K flip-flop
- 7475: 4-bit bistable latch

7490: Decade counter
7492: Divide by twelve counter
7493: 4-bit binary counter
7491: 8-bit shift register
74122: Retriggerable monostable

There are also commercial and military options to the logic devices described. Commercial devices have a narrower operating temperature range than the military. There are higher costs incurred with the extended-temperature-range devices, which do not offer any advantage in a commercial application.

7400 logic: commercial temperature range = 0°C to 70°C
5400: military temperature range = −55°C to 125°C.

Power Supply Decoupling

When any digital device changes state—and that's what, of course, they do—the power supply current will change as a result. As far as the circuit being powered is concerned, this change in current is noise. Any noise, of course, is unwanted. To get rid of the noise, the power line is decoupled by the use of a capacitor to filter out the noise and consequently stabilize the power line. A small capacitor, usually 0.1 µF in value, is used. Individual 0.1 µF capacitors are placed as close as possible to the integrated circuit itself, to ensure that switching currents emanating from the device are minimized. The changing supply currents are smoothed out by the use of a large 100 µF capacitor placed across the power line.

CHAPTER **3**

Testing Principles and Test Equipment

Before any significant evaluation of digital circuit operation can be made, there has to be an appropriate signal stimulus (the source) and a device for monitoring output transition levels (the load). This chapter describes the matching of hardware to monitoring needs. In a way, although the previous chapter has shown the digital circuit to be more difficult than the analog circuit to comprehend and, at this early stage, to appreciate its usefulness, the test and monitoring needs are, ironically, much simpler. Basic analog testing at the very least involves a multimeter, and even that allows only basic dc measurements of voltage and current. Amplifiers, our basic analog workhorses, also require, ideally, an oscilloscope and signal generator if we're to make any interrogation of an amplifier's gain and frequency-response characteristics. The inherently repetitive nature of an analog signal necessitates the higher level of sophistication required for the test and measurement equipment needs. Even a basic amplifier configuration, such as the single-stage inverting operational amplifier, needs this test array. A basic digital circuit, on the other hand, such as a NAND gate, can be monitored and verified with much less test equipment. Of course, you can use the same oscilloscope and function signal generator (or pulse generator) with the digital circuit, but it is a "nice to have" rather than a "have to have" option. The oscilloscope is by far the most versatile diagnostic equipment available for circuit verification, be it analog or digital; it's going to be your workhorse test equipment.

Test Principles

General

The only way we can really be sure of the correct operation of a circuit is by monitoring and characterizing (i.e., measuring) the output. The acid test of any electronic circuit is to verify if it is functioning according to specification. The specification could be detailed, such as performance within a certain amplitude and frequency domain, or very simple—does it work? In the case of a simple audio amplifier, if it produces sound of an acceptable quality, it's likely acceptable, regardless of its actual gain or frequency response values.

So, given that an assembled circuit is in front of you, in order to locate the faulty section of it you're going to need two things. First, you need a number of basic pieces of test equipment, typically a multimeter, oscilloscope, and signal generator or function generator. As the function generator does better as a general-purpose instrument than the signal generator, we're choosing it here. Secondly, you need to know how to go about interrogating a circuit and making sense of the test results. Test equipment, other than your basic multimeter (be it analog or digital), can be expensive; however if you find yourself working for an electronics company, where products are being designed, manufactured, and assembled, you might be fortunate enough to have access to a variety of test equipment. There are often opportunities to work on your own circuit, such as outside the normal hours. With that thought in mind, we're going to progress through the fundamental concepts of troubleshooting technique.

So, what is troubleshooting? Troubleshooting can be defined as the process of taking a nonfunctioning circuit and, through a logical process of probing, determine the cause of the malfunction. Success comes only with experience, but we've all got to start somewhere. The approach we've outlined below starts with an understanding of how circuits can be placed into a few basic categories, and from there goes on to investigate how each category can be tackled.

As analog is simpler from a design point of view, that's where we're starting. It is fortunate that when considering circuits of a fundamental nature, that is, of the type that beginners to electronics will encounter, we can broadly identify two categories, which makes the task of outlining troubleshooting easier. It goes without saying that an understanding of how a circuit (whatever it is) operates is fundamental to any study of how to find what's wrong with it.

Category 1: Circuit Block with an Input Terminal and an Output Terminal

We're going to consider circuits here as basic building blocks, not really concerning ourselves with all the components that go into those blocks but focusing on two issues. Does the circuit have an input connection, such as you'd find in a pre-amplifier, power amplifier, or guitar amplifier? If it has, it can be placed into this category. The next question (actually it is more of a conclusion) is easy: if a circuit has an input connection, it has to have an output connection also. So now we know what defines a category 1 circuit block. Think of a guitar amplifier; the input is the cord running from the electric guitar. The output is the speaker, which is usually integrated into the amplifier but still connected to the amplifier's output (the two wires on the back of the speaker have to go somewhere). The benefit of describing a circuit as a block is that for the purpose of diagnosing faulty areas, it can initially be treated merely as a device having an input connection and an output connection, without concern (at this stage) over what "goes between" the two.

Category 2: Circuit Block with an Output Terminal Only

Having defined the above configuration, we can follow the same logic and come up with a definition of a category 2 device. This is a category having only an output. What would be an example of such a device? Oscillators would fall into this category; they only have an output terminal.

Having defined these two categories, you can now identify into which of these the particular circuit you're working on falls into. Let's assume for simplicity that it's a basic operational amplifier, and let's start from there.

General Testing Procedure

The device under test is an op-amp pre-amplifier (inverting mode).

Step 1

Determine first of all that all the connections are exactly as shown on the schematic, that there are no open circuits or short circuits, that all polarity-dependent components are correctly polarized into the board, and all the solder connections are shiny and reliable. These check points might seem obvious, especially if you're checking out a circuit constructed by someone other than yourself, but it is surprising how often the simple mistakes throw a circuit into the failure. Very rarely do components fail, especially if they haven't been mistreated. If you don't do this preliminary check first, you're likely to waste an inordinate amount of time troubleshooting a circuit that has been incorrectly assembled.

Step 2

Make sure the power supply is correct. If this is a nine-volt battery, ensure that it's fresh; otherwise discard it if you're uncertain. Batteries can often appear to be correct when measured out of circuit, but under load—that is, when driving the circuit—the voltage will fall off because the battery is too old.

Step 3

Monitor the supply voltage. Measure the supply current that's drawn by the load. There's good reason for these simple confirmations. If you have more than one multimeter, so much the better. The place to locate the first voltmeter (multimeter set to the dc voltage range) is immediately after the power is fed to the circuit. This will ensure that when the power switch is turned on, there'll be voltage applied to the circuit. The second voltmeter, if you have one, is used to probe directly onto the power pin for the IC under test. If you have the same voltage there, power is at least being fed to the IC. With just one voltmeter available, just probe alternately at the points defined above. Voltage measurements are always easy to make, as the circuit does not need to be disturbed.

The power-supply current measurement requires that a break be made in the power-supply feed line. The easiest way to do this is to disconnect the connection to the side of the switch feeding the supply voltage to the load (this is the general term for the circuit) and bring out two connections that'll go to the ammeter (multimeter set to the dc current range). The positive terminal of the ammeter (the red lead) goes to the switch, and the negative ammeter terminal (the black lead) goes to the load. Typically, for a circuit having one IC, such as the operational amplifier circuit shown, you would expect to see a current of the order of a few milliamps flowing. Zero current spells trouble, and so does a current reading of hundreds of milliamps!

If you have an LED indicator in the circuit, allow for the current taken by it. You can calculate how much current is taken by the LED; the rest of the total current drain should be allocated to the circuit. Alternatively, disconnect the LED (just remove one lead); now the total current indicated on the meter is a result of the circuit only. If there's zero current, there's an open circuit somewhere. Any valid circuit is going to draw some current.

Specific Testing Procedure

We've gone through the general checks listed above. Next follows the detailed procedure, which obviously has to be referenced to a specific circuit under test.

Step 1
Verify the Vcc/2 bias voltage value. You should be getting half the supply voltage, that is, around 4.5 volts when using a nine-volt supply. A gross deviation points to the equal-value resistors being at fault, either in component values or connections.

Step 2
This step is most easily accomplished with two additional pieces of test equipment, a function generator and an oscilloscope. Couple the oscilloscope probes (from channel 2) to the output terminal (after the output coupling capacitor) of the amplifier under test; couple the other channel's probes (channel 1) to the input terminal (before the input coupling capacitor). Cables to the oscilloscope are coupled to the instrument using what are called BNC sockets. These are high-frequency sockets used with shielded cable to form connections to oscilloscopes. The opposite end of the cable is usually terminated in a pair of alligator clips (red for the live end and black for the ground end). There are two of these cables supplied with the instrument, as even the most basic of oscilloscopes have two input channels. For the type of testing

we're talking about—that is, making connection to fine circuit board traces or connections—you need something smaller. There are two options: either bring out extended wire connections from the circuit board you're testing, or use an alternate type of miniature clip, the type that is designed specifically for making connections to small component leads.

The following steps are detailed to demonstrate the process of determining the gain of the operational amplifier (the ratio of the feedback to input resistor). Turn on the power to the circuit under test (we're generally assuming here that a nine-volt battery is being used). Apply a small signal sine wave (e.g., 10-millivolt amplitude is fine, as it's a convenient value to align with the vertical screen graduations), with a frequency of 1 kHz to the input (there's nothing special about this frequency choice, it's just a convenient value). This drive signal will show up on channel 1 of the oscilloscope. Stabilize the picture and shift it to the upper vertical half of the screen, adjusting it to fit nicely between major screen divisions. The signal amplitude can be fine-tuned so that the display fits exactly between two convenient major divisions. It's just going to be used as a reference signal, and the absolute value doesn't matter at all. The output signal will go into channel 2 and once more stabilize the image. This time shift it down below the image of the first. It'll be larger, of course, in signal amplitude (as we expect), given that the amplifier is functioning correctly. Count the number of graduations the output signal extends over and multiply this number by the marking of the vertical amplifier setting. If the gain according to the circuit design is calculated to be, say, ×10, the output signal on channel 2 is going to be in the region of 100 millivolts.

If there is an output signal but it is not the level expected, the fault lies with the ratio of the gain-determining resistors. If there is no signal, move the probe from channel 1 to the other side of the input coupling capacitor. If the signal's the same, the capacitor's fine. If there's no signal, the capacitor or its associated circuitry is faulty. Follow the same technique with the output capacitor. There should be a signal on both sides of the capacitor. If there isn't, there's a wiring or component error. Measurement checks on either side of these capacitors very easily determine if the ac signals are passing through (as they should).

Test Equipment

Multimeter

The multimeter is a three-in-one instrument, performing voltage, current, and resistance measurements. Voltage and current readings are further split into dc and ac modes, but for most circuit troubleshooting purposes, we're going to be using the dc mode only, especially as the circuits shown here are all dc powered. Most multimeters these days are digital in format, in that the display is a digital numeric readout rather than the older analog type, characterized by a needle pointer.

There are subtle differences between the two types, sufficient perhaps to warrant a short description of the two variants. Most notably, the digital instrument has the benefit of higher accuracy in "readability" of the measured values. The digital is displayed without any ambiguity. When you see the numerics, you just read them off as you would read the time from a digital watch face. An analog watch face is subject to a certain uncertainty when matching the watch hand against hour markings; the digital watch face takes that uncertainty. But the digital watch display is not necessarily more accurate. The "frozen" watch display (it changes generally on the minute) gives the impression of accuracy. Remember, while the digital watch display is "frozen," time is actually moving on. If you looked at the analog face, you'd see the minute hand moving. The digital watch readout can be considered just a more convenient method for readout.

Returning to our multimeter, on the analog instrument face the pointer needle is read against an inscribed scale. To eliminate parallax error, there is sometimes a curved mirror track behind the needle. You line up your eye so that you don't see the needle image in the mirror. Obviously, if you were looking at the needle from an oblique angle, you'd see the needle's image in the mirror too. It's a nice, simple way to eliminate parallax, and it is usually found on the more expensive units. The needle is a precision feature, with a very thin profile to maximize the accuracy of interpretation. There are in fact several different scales, for the different functions (voltage, current, or resistance) being measured. So interpretation in this case takes a little care—be sure you're reading off the scale you actually want. Where the scale becomes crushed, through compression of the values, interpolation becomes even more dependent on interpretation. This is a particular advantage of the digital counterpart, where no such problems await.

So why consider analog at all? There is a good reason to go analog in one particular measurement situation. Where we're observing changing dc values, as you'd find, for example, in a recording studio, analog meters are used to monitor the signals being recorded, as the rapidly changing digital figures (from a digital meter) would be of little value—the eye cannot make any sense of a random series of changing numbers. The analog meter is the far superior medium for conveying information on the dynamics of signal peaks and lows to an observer. There is a definite use for either type of instrument; it just depends on the application. For general testing applications, you'll generally see only the digital type, mainly because of the greater convenience of readout, and also greater availability.

Voltage Measurements

Of all the multimeter measurements, voltage measurements are the simplest to make—all you really have to do is place the two test probes across the voltage to be measured. If you're using a digital instrument, you don't even have to worry about matching the polarity of the probes, as the instrument self-detects any polarity reversals. A negative sign before the numerical reading tells you the probes have been reversed, but this is of no consequence when taking a reading. As a matter of good practice, however, the correct polarity should be observed. Why? If the voltage should reverse in specific measurement applications, such as a bidirectional reversing waveform, it is important to know in which polarity the voltage is really being produced. Voltage measurements are carried out with the supply voltage switched into the circuit. The unit of voltage is volts, and most of the readings encountered will be in this range.

The range setting of the instrument needs to be tailored against the maximum voltage being measured, unless you have a digital autoranging instrument (this is a multimeter that automatically changes ranges to match the voltage being measured—as with anything, sophistication comes at a cost). In the case of a standard digital instrument, the consequence of a voltage excess is nothing more than an overload indication—you just move up to the next higher range to compensate. The analog unit's a little more critical. Depending on how quickly you notice the overload, the duration of the overload, the extent of the overload, and the rate of transition to the overload, you could have either an extremely bent (and useless) needle or, if you're fortunate, a recovered instrument when the overload is removed. We have a similar situation with voltage reversal. The digital unit will precede its numeric with a "minus" sign. With the analog, you're not so lucky. The needle will be effectively forced against the left-hand end stop. Any blockage of the ordinary mechanical movement is not good! If luck is on our side, there's no damage, but . . .

Current Measurements

All current measurements require that breaks be made in the circuit under test, as by definition and occurrence, current is a flow phenomena. You've got to break into a flow to determine its properties. When a voltage is applied to a circuit (i.e., the load), current will flow as dictated by Ohm's Law, whereby current flow is inversely proportional to the load resistance—the lower the load resistance, the higher the current flow. Raise the load resistance, and the current is reduced.

Where you make the break in the circuit, determines where the current is going to be monitored, but in the general case we're concerned with making a measurement on the supply current. So between where the positive voltage is supplied to the circuit and a point immediately after the positive supply voltage is supplied to the power-on/off switch is a good place to make the break.

Of course, there has to be some desoldering involved, and that's why current measurements take a little effort. Bring out extended wire connections onto which to connect the ammeter probes. Polarity correctness has to be observed: the positive ammeter test lead connected to the positive voltage source, and negative ammeter test lead taken to the load.

Digital instruments will, of course, function whichever way the leads are connected, but (as was the case with voltage measurements) you don't really know the current polarity if the test leads are not connected properly. It's advisable to get into the habit of having the lead polarities correct. Current measurements are made with the supply voltage switched into the circuit. The unit of current is amps, though most of the typical measurements will yield smaller current values, in the thousandth of an amp, or milliamp (mA), range.

Resistance Measurements

The multimeter is set to the resistance range for these measurements. Resistance measurements are made by directly connecting the test leads across the resistor under test. Most commonly this would be a discrete out-of-circuit component. It is rare, though not out of the question, for resistance measurements to be taken on actual circuits. Should you do this, always ensure that the power is switched off. Because of the fact that several components could be in parallel with the one component you're trying to measure, the reading you get could be misleading. Whenever there's doubt, desolder one end of the component and then make the measurement—that way there's definitely nothing shunting the component under test. Resistance values are read directly off the display. Test leads for resistance measurements can be placed either way across the resistor under test; there is no polarity dependence.

With basic instruments you've got to change range when there's an overload signal. But with the analog meter, there is no damage potential. This is the most rarely used mode for troubleshooting; however one instance where resistance measurements are sometimes made is directly across the power and ground lines, with the power off, on the down side of the power switch. A direct short (zero resistance) reading will tell you that something is drastically amiss—in which case the supply voltage will sink dramatically, and the current will be excessively high. Resistor verification, especially where the color code is hard to read, is a better example of the particular use for the ohmmeter.

For an excellent example of the advantage of the analog over the digital meter, consider the following situation. Potentiometers (as we're talking about resistors) are sometimes, through wear and just rogue devices, prone to glitches in the internal resistor track. To see this visually, hook up to the ohmmeter two of the potentiometer terminals, the center terminal and either one of the outer ones. As you rotate the control knob (and we're assuming you've got a defective device here), you'll see the resistance change smoothly. When the rogue spot is crossed over, the resistance will jump; you can see this very clearly. If a digital meter were to be used instead, such a glitch would be almost impossible to discern. There's certainly something to be said for less than state-of-the-art technology!

Oscilloscope

Introduction

The oscilloscope is not unlike a regular TV receiver, which intercepts, via the antenna, incoming electronic signals (transmitted through the airwaves from the TV station) and, after a certain amount of electronic processing, produces the TV pictures you see on your screen. In a similar fashion, the oscilloscope produces an image of electrical signals that are fed into the instrument, not via an antenna but directly through connections on the front panel. The only reason for the TV antenna is to capture the wireless signals. In principle, the transmission signal could be directly coupled into the TV receiver, if, for example, we had direct access to the transmitted signal. The oscilloscope is unique in its ability to produce a visual image of the variation over time of an electrical signal, unlike, say, the multimeter, which is able only to display the average magnitude of such a signal. How does the oscilloscope perform this feat?

Basic Principle of Operation

This display-over-time signal on the oscilloscope is accomplished by repeatedly sweeping (electronically) the signal across the face of the oscilloscope screen. But we require more than just a "sweep." There are two phenomena related specifically to the operation of the instrument. First, there has to be a suitable physical medium onto which whatever is imaged is displayed. This medium is called a phosphor screen, which has the property of fluorescence, that is, it glows visibly when struck with an electron beam. The second is the persistence-of-vision property of the human eye. This can be explained as retention by the eye (for a short period) of an image of an object being viewed after that object is removed from sight. If we therefore were to see this retained image repeatedly (by some means), we would start to see these image-flashes as if they were a stationary image, not a rapidly repeated series of the same image. These two properties enable the oscilloscope's operation.

Consider any electronic signal that varies over time, such as a square wave that pulses between two voltage extremities, a high level and a low level (this is the definition of a digital signal). We need a mechanism whereby this action can be duplicated on the oscilloscope screen. This is achieved by moving an electron beam across the screen (from left to right) at the same speed (or a multiple of the same speed) at which the pulse waveform is repeating.

Don't forget, this is a repetitive waveform we're using; it has to be repetitive, i.e., continually going through cycles and starting again. A waveform cycle is defined as the time interval between any reference point in the waveform and when that point is reached again. If we start with the reference point where the lowest voltage level makes the transition to the maximum voltage level, the square wave will stay at that positive maximum, drop to a minimum value, rise to maximum once more and finally drop to the minimum again. It'll then repeat again. At that point we have what is defined as one cycle of the waveform.

We want to move the electron spot across the screen at the same speed as the waveform, so that when the spot reaches the far right-hand side of the screen, it immediately shoots back to the left-hand side and starts moving again. This direction of the spot's movement (from left to right) is defined as being along the x-axis. But if this is all we did, all we'd get is a straight line across the screen, of hardly any use at all! So to get more information, we need also to move the electron beam vertically—defined as the y-axis—again by an amount that corresponds to the magnitude of the pulse signal (or a multiple of it). Now when we combine this vertical action with the spot's sweeping horizontal motion along the x-axis, there magically appears an image of our starting pulse waveform. What we are seeing is essentially a display over time of the signal's fluctuations. Once the electron beam has reached the right-hand edge of the screen, it rapidly (too rapidly for the eye to see) returns to the starting position on the left-hand side, and the sweeping action starts again.

Now, because of the eye's persistence-of-vision property, we start to see a stable image, if the repetitive sweep rate is sufficiently fast. And yes, as you've expected, there is a front panel control on the oscilloscope that governs the sweep rate, covering a very wide range to handle the full spectrum of both fast-moving and slow-moving signals. If we slow down the sweep rate, you can see the images appearing as separate flashes. The input signal is fed into the input-amplifier—also called y-amplifier, as the signal is being intrinsically displayed in the vertical, or y, direction. Without the sweep applied (and it can be turned off), all you'd see is a straight vertical line, the height of which would depend on the level to which the input signal is amplified. Even budget-priced oscilloscopes come equipped with two of these y-amplifiers, thus allowing two signals to be displayed at the same time. This is a very nice feature to have when you want to look simultaneously at both the input and output signals from an amplifier. By measuring the relative amplitudes of the output and input signals, you can determine the gain of the amplifier stage.

Oscilloscope Controls

Gain The gain controls of the y-amplifiers control the vertical size of the signal you see displayed on the screen. These controls are calibrated in units of volts per division, where "division" refers to the major grid spacing marked on a plastic graticule that is overlaid on the face of the oscilloscope. Assuming you've got a one-volt signal feeding into the y-amplifier and the y-amplifier gain control is set to 1 v/div, you're going to get a signal displayed that's one division high. A division is a little too small to see any detail; how do we get a larger picture? Easy—just change the setting to a more sensitive setting of, say, 0.5 v/div, and you get a signal that's two divisions high. Typically on an oscilloscope you would find eight divisions covering the vertical expanse of the screen, thus enabling a wide range of signal amplitudes to be displayed (from millivolts to tens of volts in amplitude). The markings for the y-amplifier control vary in steps from multiples of volts/div to millivolts/div.

Sweep Timebase The controlling circuitry for the sweep voltage that swings the signal across the screen from left to right is called a timebase. The timebase control is similar to the y-amplifier gain control and has graduation marks in units of time/div. This control covers a range of sweep rates from "slow" (seconds/div) to "fast" (microseconds/div). As the sweep rate control is rotated clockwise (to cover divisions in shorter times), the sweep rates increase, i.e., the electron beam moves faster across the screen. The timebase control is a versatile means of getting a good estimate of the frequency of an input signal. It's an estimate, because we're trying to gauge a numerical value from a visual measurement. Nevertheless, it's an excellent method of making a quick frequency calculation.

This is how the measurement is made. Assume that we have a repetitive signal, such as a sine wave (let's assume it's a 1000 Hz frequency for convenience), displayed in the screen. The controls have been adjusted to get a nice, stable image, and we have one or two cycles displayed across the screen. The objective of the measurement is to locate two similar, or "like," points on the signal. The sine wave is easier to effect a frequency measurement on than a square wave, and it is also more closely associated with the concept of frequency than is a square wave. We can choose, for example, two consecutive positive peaks and measure the distance between them, by counting the number of horizontal divisions in that interval. If the timebase setting is, say, 1 ms/div (one millisecond/div) and there is one full horizontal division between the peaks, this is equivalent to 1 ms; this distance is defined as the period of the signal. To find the frequency, we merely take the reciprocal of this number; 1/1 ms is 1000/1 Hz = 1000 Hz. That's how simple it is. The accuracy of the measurement depends on how accurate you can estimate by eye. Because of the effect of parallax, which is the error caused by not viewing a signal against the graticule squarely on, there could be loss of exactness in the calculation we make. There is also the compounding error of thickness of the image line and of the measuring lines on the graticule.

Once you're familiar with the sine wave, you can work the same magic with a square wave and measure the distance between two consecutive similar points on the waveform, e.g., from the point where the pulse edge starts to rise to where it does this again. For both sine waves and pulses, a cycle period for the wave shape can be defined as the period between where the signal starts at zero and where, having risen to a positive maximum, dropped to and past zero, and fallen to a negative minimum, returns again to zero. Waveforms that are periodic in nature—that is, they follow this repeating pattern forever—are the most common types encountered when making measurements using oscilloscopes. In order for signals to be displayed they must be repetitive. Signals that are not, and these do exist (they're either transient signals or nonperiodic signals), require a different type of oscilloscope.

Brightness and Focus Secondary controls on the oscilloscope allow the user to vary the display brightness and sharpen up the focus. Excessive brightness is not necessary and can cause the image to burn into the screen phosphor, so keep the brightness down to a just-sufficient low level.

Vertical and Horizontal Shift There are also two controls for moving the trace either horizontally or vertically. These have bidirectional arrow markings on them, indicating which way the displayed trace is capable of being shifted. A calibration signal source (a square wave with equal on and off periods) is always provided on the oscilloscope. If this is, say, a 1 kHz, 0.2 Vpp (peak-to-peak voltage) calibration source, coupling this signal into one of the y-amplifier inputs will display a signal that is one division high with one waveform cycle covering two divisions, when the y-amplifier control is set to 0.2 V/div and the time base set to 0.5 ms/div. The purpose of having this calibration signal there is, of course, to check the calibration of the vertical amplifier and horizontal timebase controls.

dc/ac Coupling Signals can be coupled in either a dc or an ac mode to the y-amplifier. The actual difference between these two modes is whether a capacitor is used at the input for coupling or there is a direct coupling into the y-amplifier. The presence of the capacitor will remove the dc component from any waveform. If we were to take a square-wave signal of 0.2 Vpp and couple this in a dc mode, we'd see the signal positioned on the zero volts reference line and making an excursion to a positive value of 0.2 V. With the same signal applied, but now coupled in an ac mode, the display will be shifted downwards so that the waveform is positioned equally above and below the zero volts reference line (this is only relevant if the waveform has equal on and off periods). This difference in the display is not significant for the two coupling modes, although, of course, the calibration has to be done in the dc mode.

A more significant example of the use of the ac mode is seen in the following. Let's say you're displaying a large-amplitude dc signal, and mixed in with this dc signal is a small amount of noise. If you were to increase the gain by increasing the input sensitivity of the y-amplifier, in an attempt to amplify and observe the noise, the dc signal would also be amplified. When this happens the trace gets too large and goes off the screen. This, of course, doesn't help matters. But by switching to the ac mode, we will now remove the dc component of the signal, and only the noise (which is the ac component) will be amplified as the gain setting is increased.

Trigger Control This feature on the oscilloscope is the mechanism by which repetitive signals can be frozen onto the screen and appear as a stationary image when in fact they are in a state of constant undulation or flux (like, of course, an ac signal). There are two variables within the user's control for stabilizing, freezing, or locking the display; these are the threshold voltage and the polarity of the slope of the waveform. As an illustration of how to achieve a stable waveform, assume again we have an input feed of a sine-wave signal. Start by setting the threshold level control to the midpoint position and the slope switch to the positive value. These are controls on the oscilloscope front panel. As the threshold level control is slowly adjusted to the positive value, you'll see the effect on the trace: the waveform will start to trigger later (than the zero crossing point). When the threshold control is rotated to the negative value, the waveform will start to trigger earlier (than the zero crossing point). So, basically, what this control does is to vary the start point of the waveform. If the control is set to a value that is outside the range of the waveform, the lock will be lost, and the waveform will become an unstable, running image.

Slope Selector The second control we have is the slope selector switch, which allows the user to choose to start the displayed waveform at either the rising section (positive slope) or the falling section (negative slope) of the input waveform.

External Trigger Another useful control is the external trigger control, which allows weak signals to be more positively locked. This is done by coupling the signal under examination into the external trigger socket also and selecting the external-trigger source switch option.

Function Generator

Introduction
The function generator is a very versatile test generator providing the stimulus signals necessary to test most of the basic circuit projects encountered here. The function generator produces three distinct waveforms—square wave, sine wave, and triangle wave—which makes it versatile when testing circuits. A signal generator produces high-purity sine waves, but from a versatility point of view the function generator wins. Its operation is really simple: select the type of output you want, and that's basically it. The easiest way to see your output is to couple the function generator output directly into an oscilloscope. Nothing could be simpler. As the function generator also provides a useful sine-wave output, there is no need for a separate sine wave signal generator. For most general-purpose applications, the sine wave produced by a function generator is more than adequate.

The onset of distortion in audio amplifiers through signal overload can be easily observed using such a sine-wave input to the amplifier under test and cranking up the amp's gain until overload starts to occur. Usually we don't want this to happen, unless there's a specific requirement for it, as in the popular "overdrive" function on guitar amplifiers, also known as "distortion" or "fuzz" controls. Square waves are a quick and easy way to get an indication of an amplifier's frequency response. The triangular wave is similar to the sine wave in shape and can be used for gain testing.

Basically that's all there is to the function generator. A clearly marked BNC connector (a special type of connector designed for radio-frequency signal use) is on the front panel. Plug in a BNC-terminated shielded cable between the function generator output and the oscilloscope's input and switch on, turn up the signal amplitude, and you've got a waveform on the display.

Function Generator Controls

Amplitude A commercial function generator is designed to work into a low-impedance load of 50 ohms (this is a common industrial specification), so you'll often see specifications listing the maximum output voltage as something like 20 Vpp (peak to peak) into an open circuit or 10 Vpp into a 50-ohm load; the actual values, of course, depend on the individual instruments. Commercial circuit designs are often (where this is a design requirement) configured to have specific input and output impedances. This is so that the performance of the unit when interfaced into other equipment can be guaranteed. Hobbyist designs rarely fall into this category, as interfacing with guaranteed operating specifications is not a concern.

The function generator output voltage is more likely to be used at a minimum voltage setting, as it's a feed instrument, in that it supplies a low-level signal that is used as a source of stimulation. A built-in attenuator will take the minimum output to a value that's as low as a few tens of millivolts again into 50 ohms.

dc Offset A dc offset control, often provided, enables a ±10 volt swing about the zero value. For general ac applications you won't need this feature, as the coupling capacitors in ac amplifiers remove any such dc component.

Frequency Range The function generator has pushbutton range selection switches (typically eight) that allow overlapping frequency ranges, usually from 1 Hz to 10 MHz, in eight steps to be selected. Each selected frequency range works in conjunction with a continuously variable frequency control that allows any frequency between 1 Hz and 10 MHz to be accessed.

Duty Cycle A variable-duty-cycle option enables duty-cycle ratios on either side of a 50 percent value to be obtained.

TTL/CMOS Depending on the individual manufacturer's instruments; interfacing to popular TTL or CMOS digital logic circuits is taken care of with a switch option that allows supplying either type of matching signals. TTL (transistor-transistor logic) circuits operate from a different supply voltage (typically five volts) than CMOS (complementary metal-oxide semiconductor) circuits (typically 15 volts), hence the two selection options. But be aware that some CMOS circuits operate off the same five volts as TTL, so choose the range accordingly. TTL circuits are by far, though, the more popular type of digital circuit, and they are taken of with a fixed +5-volt output. This is an important feature, as it automatically ensures that whatever waveform output is selected will be no more than five volts in amplitude.

Frequency Sweep A special feature of a function generator (unlike a signal generator) is its ability to produce a swept waveform—that is, a nominally selected center frequency can be swept through a user-defined range of frequencies. The swept frequency repeats continually through the range selected.

Project 1: LM 555 Demo Oscillator

Introduction

The demonstration circuit provided below can be used as a test bed illustrating the use of test instruments.

Circuit Description

This is a very simple, free-running, square-wave oscillator, based on the popular LM 555 integrated circuit (IC1) and configured to run at a very slow rate so that the transitions can be easily seen by the human eye. The schematic is shown in Figure 3-1. Apart from the IC itself, there are only four components (two resistors and two capacitors) needed to generate the train of square waves. It is so simple that the construction could well be done within an hour. Power is supplied by a nine-volt battery, and the circuit will continue to function even if the battery voltage falls to five volts. Additionally, as the current drawn from this circuit is so small (around 5 mA), we can expect a very long life from the battery. The timing part of the circuit is controlled by the two resistors, R1 (10 kohm) and R2 (100 kohm), and the large electrolytic capacitor, C1 (22 µF). Basically, the higher the timing component values, the lower the frequency the output will be—hence the unusually high capacitor value of 22 µF.

The remaining capacitor, C2 (0.01 µF), is needed to get the LM 555 operating in the correct basic mode. Once the power is applied and the device is fired up, you'll see that the output is derived from pin #3 of the LM 555. The waveform immediately available from pin #3 is a positively going waveform—that is, it starts from the zero voltage as the baseline and rises to a positive maximum that is close to the supply voltage. It is important to remember this polarity consideration, as it dictates what we need to use as a monitor circuit.

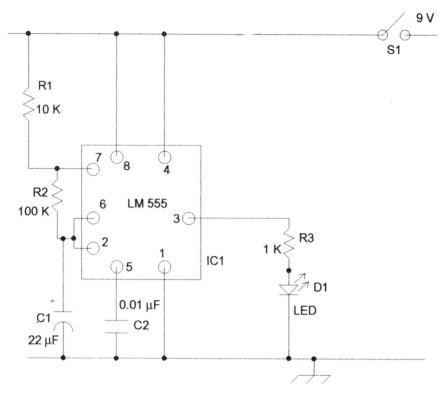

Figure 3-1 LM 555 Demo Oscillator

The portion of the output waveform that sits for a period at the high positive voltage value is defined as the "on" period. This would correspond to the high ("1") state for any logic circuit. Of course, the voltage will have to fall to zero again at some point; the period during which the waveform sits at the zero voltage level is defined as the "off" period. This would correspond to the low ("0") state for any logic circuit. The "on" and "off" periods are measured in units of time (seconds), whereas the signal amplitude is measured in units of volts. The sum of the on and off periods is the total period, which defines one cycle of the waveform. The ratio of the "on" to the total period is defined as the duty cycle. In the special case where on and off periods are equal, the duty cycle has a value of 50 percent. A duty-cycle value greater than 50 percent signifies that the on period is greater than the off period. Pulse waveforms, of which the square wave is one, are often described in terms of duty-cycle value.

Signal Monitoring

A first check of the circuit to verify the circuit is functioning is made simply by connecting a dc voltmeter or multimeter, set to the dc voltage range, to measure a dc voltage in the region of nine volts or so. An analog multimeter is preferable to a digital multimeter, as the voltage excursion is much more evident. You should see (given that the circuit has been correctly assembled) the meter needle oscillate between a positive voltage value and a zero voltage value. If you see this, the circuit is operating correctly.

The second part of the circuit deals with the monitor. Switch off the power and wire in the extra two components, as shown in the schematic. This is just a basic LED (light-emitting diode) with current-limiting resistor. Depending on the way the LED is connected in the circuit, it'll turn on whenever a positive voltage is supplied to it. The test waveform circuit we've just built is producing just that. So when you now turn the power back on, you'll see the LED flash each time the device goes into the on state—simple as that. When the pulse falls into the off state, the LED goes out. When the transitions between the on and off states are slow enough for the eye to register, you can very roughly make out the duty cycle, by estimating or counting (in time) the relative durations of the on and the off periods.

So, as you can see, where the digital circuit has a waveform that changes slowly enough, a simple LED is all it takes to act as a pulse monitor. You couldn't get simpler than that! Note that as the frequency increases, the LED pulses faster, until the eye perceives it as being constantly on. After that point, we can't be sure if the circuit is actually functioning or is locked in a malfunctioning "on" state all the time. Where can you, with an analog circuit, find such a simple output state verifier!

Oscilloscope Measurements

With the source and monitor sections of the circuit now established to be correctly operating, we can use the oscilloscope actually to determine waveform values. I'm assuming you have access to an oscilloscope, and you might, especially if you work in an electronics facility. Set up the oscilloscope first, making sure the calibration waveform on the front panel is displayed solidly and that the amplitude and frequency match with the specified value (this would be in the operating manual). Hook up the oscilloscope probes to monitor the output from pin #3 and ground, then set the input to the dc mode. With a 1 v/div setting you should be seeing the spot trace up to about five divisions and then fall back to zero. Each time the LED flashes "on." the screen spot will stay at the five-volt level; when the LED is off, the spot drops to zero voltage. The maximum positive voltage reached by the square wave can be easily calculated by reading off the number of vertical divisions covered by the trace and multiplying this by the calibration setting (of 1 v/div).

Because of the slow trace characteristics (remember, this was done to allow the separation between the LED flashes to be discernible to the eye), the waveform is too slow to resemble a regular square wave. We can easily rectify this by stepping up the signal frequency. The timing capacitor has a 22 μF value. Replace this with a value about a tenth the size, that is, 2.2 μF. The waveform can be now seen to have a smaller period (but not exactly a tenth of the previous one, as the period is function of both the capacitor and resistor values). You can get a better indication of the square-wave form.

For an even "better-looking" waveshape, the capacitor can be reduced again to 0.22 µF to get a higher repetition rate (that's another way of saying it's a higher-frequency square wave). As you increase the frequency, the timebase can be incremented clockwise to get a stable screen image, making the determination of the waveform's period much simpler. Of course, at this stage the LED will appear to be constantly on (if it hasn't already with the previous capacitor change). Of course it's pulsing, but the eye cannot discern the changes, so it appears always "on." The aim of this exercise is to gain familiarity with using LEDs and oscilloscopes as output monitors. Each method serves its own purpose.

In case you're interested, though, the calculation for the frequency of oscillation is given by:

$$f(Hz) = 1.44/[R1 + (2 \times R2)] \times C$$

where

f (in hertz) is output frequency
R1 (in ohms) is the resistor going from pin #7 to Vcc
R2 (in ohms) is the resistor between pin #7 and #6
C (in farads) is the capacitor between pin #6 and ground.

This circuit is extracted from a more comprehensive circuit shown later in chapter 5 (Project 2); it is placed here as a demonstration test vehicle. However, actual component values are shown below, in case you want to construct this circuit. It is a real, working circuit in the format shown in Figure 3-1, with my customary power-supply section components omitted. This has no impact on its operational viability.

Parts List

Semiconductor
    IC1: LM 555 timer
Resistors (all resistors are 5 percent, 1/4 W)
    R1: 10 kohm
    R2: 100 kohm
    R3: 1 kohm
Capacitors
(All nonpolarized capacitors disc ceramic)
(All electrolytic capacitors have a 25 V rating)
    C1: 22 µF
    C2: 0.01 µF
Additional materials
    D1: LED
    S1: single pole single throw miniature switch
    Power supply: nine-volt battery

CHAPTER **4**

Digital Circuit Truth Tables

Truth tables for the common building-block gates (AND, OR, NAND, NOR, EX OR, EX NOR) are introduced in this chapter. Understanding logic circuit operation involving more than one input (which is generally the case) requires knowing the expected state of the output. It is this outcome that is described in truth tables. Depending on the type of logic gate used, for example, AND, NAND, etc., the output logic state (in response to a specific input-state stimulus) will vary. Hence, knowing the truth table associated with that particular type of device is critical. Details are provided for the input/output states for the six basic gate types listed above.

There are four basic logic blocks that are the cornerstone to all logic design. Let's take a gentle look at these. The first and easiest logic block we encounter is the AND gate. From the introduction, we expect this AND gate to take the form of a two-terminal input device with one output, and that is exactly what it is.

AND Gate

The AND gate is a logic device where the output will only go into a logic high state when both of the inputs are at a logic high; all other conditions will produce a logic low (0). For simplicity we are using here as an example a simple two-input logic gate, having an input A, input B, and output C. A logic low at the output will occur only if any one of the inputs is at a logic low. The standard way of representing such a logic combination is by what is called a "truth table." A truth table merely lays out in columns the input and output states. For a two-input device, the truth table looks like the table below.

AND Gate Truth Table

Logic State	Logic State	Logic State
a) If input A = logic 0	AND input B = logic 0	then output = logic 0
b) If input A = logic 0	AND input B = logic 1	then output = logic 0
c) If input A = logic 1	AND input B = logic 0	then output = logic 0
d) If input A = logic 1	AND input B = logic 1	then output = logic 1

If we read across the rows of this truth table for a two-input AND gate, it says that whenever both of the inputs are of the same state—that is, when the inputs are either both logic 0 or both logic 1—the output will follow, or track, that same state. So we can see that when both inputs are at logic 0, the output is at logic 0 too (as in row a). When both the inputs are at logic 1, the output is at logic 1 too (as in row d). But notice that the output remains in the low logic 0 state if the two inputs are different in logic values (as in rows b and c). At this stage you might be wondering what the use of this AND gate might be. Like everything else, we have to accumulate a certain threshold of knowledge before things become clearer. So be patient—the more interesting logic circuit builds come later.

Let's say we had the equation $X + Y = Z$. Z is always given by the sum of the first two components.

If X were 0 and Y were 0, we'd get $Z = 0$.
If X were 0 and Y were 1, we'd get $Z = 1$.
If X were 1 and Y were 0, we'd get $Z = 1$.
If X were 1 and Y were 1, we'd get $Z = 1$.

Notice how different this result is from the above, because this is an algebraic relationship we're dealing with here; it's a straightforward summing of two values to produce a third.

The AND gate rule, though, is based on logic and goes like this. The output from an AND logic gate will go into a logic high state (1) when all of the inputs are at a logic high state (1). Conversely the output will go to a logic low state (0) when any of the inputs are at a logic low state (0).

This definition becomes clearer if we consider the sequence of events leading up to these two states that the output can take. Let's start with a case where both of the inputs are low. In reality, you can do this by grounding both inputs. If you were to do this and to act as a monitor on the output, on which we place a voltmeter (set to read dc voltage), you'd see the output voltage read zero volts. We are assuming that the device is properly powered up to begin with. If this were a common logic type of device called a TTL (transistor-transistor logic), the power supply would be +5 volts. (TTL is a relatively high-speed logic device based on bipolar technology, with a medium power dissipation. It's inexpensive and forms the basis of most of the commonly found logic designs.)

But going back to our sequence of events, let's say that at time T1 we have the situation shown below. Then at time T2, we take one of the inputs high, by connecting, say, input B to the +5-volt rail. What will happen to the output? Nothing, according to our truth table, since by definition of an AND gate, we need both inputs to go high before the output will go high. Now, at time T3 we swap over the logic states to the two inputs and make input A high by connecting it to the +5-volt rail, and make input B low by connecting it to the ground rail. Should we see any difference to the output? Again there is none, as dictated by our AND truth table. At the next time interval, T4, we keep the previous input A high and this time also take the other input B high. The output, C, will immediately go high, as seen by our dc voltmeter's giving a positive reading. Next, at time T5, we once more take the B input to a low state. The output immediately falls to zero, that is, there's a change of state at the output, reflecting that particular change at the input. Finally, at time T6 we also take the input A to a low state. There is no change in state in the output for this condition, as the output was already low.

AND Gate Truth Table Event Sequence

Time	Input A	Input B	Input C
T1	0	0	0/starting value
T2	0	1	0/no change of state
T3	1	0	0/no change of state
T4	1	1	1/change of state
T5	1	0	0/change of state
T6	0	0	0/no change of state

As you can see, with a simple digital logic gate the output will change state depending on where it was before the input changed state, and also on the type of change of state the input was going through. Obviously this is a little more convoluted than a basic analog operational amplifier, which to all intents and purposes can be described much more simply like this: with a ×10 gain, a 1 mV input will generate a 10 mV output signal, regardless of how and when the sequence of operation takes place. We've got to be a little more careful with digital circuits. Our starting point could be anywhere in the sequence for the AND gate; if we started with the T4 condition, where both inputs were high, the output would correspondingly be high. We could go forward to the T5 condition, and the output would change state. Alternatively, we could go directly to the T6 state, and the output will still change state. In fact, if you take a close look at the column for output, you'll see that if we start with the T4 state, whatever change we make to the input, whether this be T1, T2, T3, T5, or T6, the output will always change state. Taking the scenario a stage farther, if we had instead started with, say, the T3 condition, the only situation that will cause a change in the output is going to the T4 state. Any other change, such as going to the T1, T2, T5, or T6 states, will not cause the output to change at all.

As you can see, there are a lot of interesting observations to be gained from the truth table of a simple logic AND gate. With a supply voltage of +5 volts for a TTL AND gate, the output high state is typically specified by the manufacturer to be +3.5 volts and for the low state 0.2 volts, which is close enough to zero volts. When taking an input to a high level, we have to exceed at least +2 volts. Similarly, when taking an input low, the voltage must be less than 0.8 volts. These voltage conventions are the norm with TTL circuits. A high voltage could be Vcc, and a low voltage could be ground, so in the example we've been using for the AND gate truth table, we could substitute those voltage levels instead. Note that the power-supply requirements for TTL logic are very precise and have to be within a narrow range of +4.5 to +5.5 volts. Analog circuits, such as the basic operational amplifier, are much more tolerant of supply voltage variations. In practice, with digital TTL logic, we'd use a regulator to fix the supply voltage at five volts. In essence, the regulator (which is often supplied by a simple three-terminal integrated circuit) takes a wide-ranging dc input voltage and produces a very stable five-volt output, which is used to drive TTL circuits directly.

AND Gate Truth Table Sequence (Voltage Mode)

Time	Input A	Input B	Output C
T1	0 V	0 V	0.2 V/starting value
T2	0 V	5 V	0.2 V/no change of state
T3	5 V	0 V	0.2 V/no change of state
T4	5 V	5 V	3.5 V/change of state
T5	5 V	0 V	0.2 V/change of state
T6	0 V	0 V	0.2 V/no change of state

Another way of interpreting the AND gate is from an electrical viewpoint. Here, in this simple arrangement, we have two switches A and B located in series with a battery and an LED (with resistor). When switch A is closed, what effect does this have on the circuit? None, as the LED can't be lit, because there is still an open circuit through switch B. Alternatively if switch A were left open and switch B closed, the result would be the same: the LED would not be lit. In order to get the LED lit, both switches A and B would have to be closed. That's exactly the state of affairs we have with the AND gate truth table. So you can see, using an electrical analogy makes understanding it simpler.

OR Gate

The OR gate is a logic device in which the output goes high when one or both of the inputs are high (1); when both of the input are low, the output is low (0). It (the OR gate) also covers the either/OR state, as seen in condition d) from the table, that is, when either input A or input B is high (1), the output C will go high (1).

For simplicity we are using here as an example a simple two-input logic gate. A logic zero at the output is produced only when all of its inputs are at a logic zero. In this logic gate, we again have two inputs A and B, and an output C. We're going through the same stress sequence of changing the input terminal levels to either a high or a low state, as we did before for the AND gate. For a two-input device, the truth table is:

OR Gate Truth Table

Logic State	Logic State	Logic State
a) If input A = logic 0	OR input B = logic 0	then output = logic 0
b) If input A = logic 0	OR input B = logic 1	then output = logic 1
c) If input A = logic 1	OR input B = logic 0	then output = logic 1
d) If input A = logic 1	OR input B = logic 1	then output = logic 1

Specifically as the OR title implies, the output will switch states and go into a high state (from an initial low state) if either of the inputs (A or B) is taken high, but note also that the output will make the same transition if both inputs go high. The OR gate does not distinguish between whether one or both of the inputs are high; this is a logic rule with the truth table, and it has to be learned or remembered.

If we run through a time-elapsed sequence of events in the example below, we can see when changes to the output would occur. Digital logic is designed having principally in mind the conditions under which the output changes; it always has to be with respect to a starting position. In actual logic circuits, the design ensures that when the circuit is switched on the same initial logic condition is always established, so that the subsequent logic transitions can be guaranteed upon application of an input stimulus. This is critical point, since if we didn't know what the starting conditions were, we wouldn't be able to determine into what logic state the output would move.

So, let's say at time T1 we have the two inputs at a logic low (0). This being our starting state, the output would also be at a logic low (0). At the next time interval, we take input B to a logic high (1). From the OR truth table, we know that the output will switch states into a logic high (1). This is the situation shown at time T2. If we were to make the transition shown as T3, where the input situations were reversed—input A is taken high (1) and input B taken low (0)—the output would remain at the previous high (1) state.

Incidentally, if you're trying to do this experimentally with simple switches to toggle or switch the inputs, be aware that you are unlikely to get the indicated output states. The reason is that mechanical switches produce voltage transients as the switch contacts open or close. In doing so, the logic state (high or low) being applied to the input is unknown or indeterminable, as no clean pulse is being applied; hence, the output in turn can take on some unknown state. Switching should be done electronically, as you'll see later, in order to guarantee a clean pulse; that way you'll get the correct logic output states as shown below.

But back to our table. We had left off at the T3 position, where the output state was at a logic high (1). The next time interval finds us with input B also taken high (1). The consequence of this action on the output is nil, as the output state remains at a logic high (1). Following the same pattern as before, the time T5 situation is input A high (1) and input B low (0). The output remains at the high (1) state, until finally at time T6, when both inputs are taken low (0); the output finally switches state to low (0).

OR Gate Truth Table Event Sequence

Time	Input A	Input B	Output C
T1	0	0	0/starting value
T2	0	1	1/change of state
T3	1	0	1/no change of state
T4	1	1	1/no change of state
T5	1	0	1/no change of state
T6	0	0	0/change of state

Let's consider the electrical representation of the OR gate. Our usual battery is the source, and the LED plus a resistor is the load. The intermediate switching arrangement to represent the OR logic function is this time the parallel switching arrangement of S1 and S2. Switch S1 can represent input A, and switch S2 can represent input B. We'll start off with both switches in the off position; obviously there is no power supplied to the LED, so it is unlit. Now close switch S1. Power is now available, and the LED lights up. That's the OR function: either S1 or S2 will turn the LED on. Now switch S1 off and switch S2 on; the LED is still lit, for the same reason—power is still available. If S1 is now also placed in the on position, there is no change to the LED; it's still on. Provided one of the switches is on, the action of the other is redundant.

NAND (NOT-AND) Gate

The NAND gate is a logic device in which the output goes high (1) when both of the inputs are not high (1). When both of the inputs go high, the output is low (0). For simplicity we are using here as an example a simple two-input logic gate. But the explanation is valid also for a NAND gate with more than just two inputs. The NAND gate is commonly found in the dual 4-input configuration (this is the LS20 device). The NAND gate can be considered as a combination of NOT & AND, that is, a NOT-AND gate. Because of the additional complexity of dealing with this convoluted bi-logic function, we'll describe the truth table in two ways. For a two-input device, the truth table looks like this. We can see quite clearly how the output behaves when stimulated with the input combinations. The input sequence increments as 00, 01, 10, 11, in typical fashion for a two-bit binary device.

NAND Gate Truth Table

Logic State	Logic State	Logic State
1) input A: 0	input B: 0	output: 1
2) input A: 0	input B: 1	output: 1
3) input A: 1	input B: 0	output: 1
4) input A: 1	input B: 1	output: 0

Next we spell out the sequence of events we're describing in the truth table—and this really is important, because to understand the logic table, you've got to comprehend the meaning of the table. The truth table above is a neat summary of the NAND function. This "shortened" table is how you'd see it in a data sheet. The version below is the same, except stated a little differently; actually it's more "wordy," but it does illustrate the need to comprehend exactly what the truth table states. If you actually "speak" the sentences, you'll see the table starts to make sense. Both versions of the truth table serve unique purposes for understanding the NAND gate. The NOT function takes a little getting used to, if you haven't come across it before.

NAND Gate Verbalized Truth Table

a) If input A is a logic 0 AND input B NOT a logic 1
   (i.e., it is a logic 0), the output C is a logic 1.
b) If input A is a logic 0 AND input B is NOT a logic 0
   (i.e., it is a logic 1), the output C is a logic 1.
c) If input A is a logic 1 AND input B is NOT a logic 1
   (i.e., it is a logic 0), the output C is a logic 1.
d) If input A is a logic 1 AND input B is NOT a logic 0
   (i.e., it is a logic 1), the output C is a logic 0.

The NAND gate is a little different logically from the previous (easier to follow) two gate forms (AND & OR), as it is a combination of a NOT and an AND logic function. This gate produces a high (1) output state only when both of the inputs are not high (1). It (i.e., the NAND gate) can also be defined as the complement of the AND gate, in that the output condition is the exact opposite of the AND gate truth table. It is, in fact, the most common logic gate encountered, and starting with the explanation of the "easier" AND gate makes the NAND gate simpler to comprehend.

The flow sequence as the two-bit logic increments shows when the states change. Assume that at time T1 the two inputs are in a logic low (0) state. This is our starting state, and the output under these conditions is at a logic high (1). At the next time interval, T2, input B is taken to a logic high (1). The output remains constant at the previous logic high state (1). At the next time interval, T3, the input states are swapped, with input A taken to a logic high (1) state and input B to a logic low (0) state. The output again is at a logic high (1), with no change in state. At time T4, both inputs are sitting at a logic high (1), and the output finally switches into a logic low (0) state. As we sequence the inputs down once more, the next time interval, T5, sees the input A at a logic high (1) and input B at a logic low (0). The output switches state once more to a logic low (0). Finally, as we reach time T6, both inputs are again at a logic low (0). The output remains unchanged sitting at a logic high (1).

NAND Gate Truth Table Event Sequence

Time	Input A	Input B	Output C
T1	0	0	1/starting value
T2	0	1	1/no change of state
T3	1	0	1/no change of state
T4	1	1	0/change of state
T5	1	0	1/change of state
T6	0	0	1/no change of state

Another way to look at the NAND gate is to consider it as the negation of the AND function. The sequence below steps through the logic transitions for this interpretation.

Step 1

Let's start with the AND function as shown below. Set up the AND gate truth table as shown.

Logic State	Logic State	Logic State
a) If input A = logic 0	AND input B = logic 0	then output C = logic 0
b) If input A = logic 0	AND input B = logic 1	then output C = logic 0
c) If input A = logic 1	AND input B = logic 0	then output C = logic 0
d) If input A = logic 1	AND input B = logic 1	then output C = logic 1

Step 2

Next negate the AND function by changing the title to NOT-AND and accordingly inverting the output states, so we get:

NOT-AND Gate Truth Table

Logic State	Logic State	Logic State
a) If input A = logic 0	AND input B = logic 0	then output C = logic 1
b) If input A = logic 0	AND input B = logic 1	then output C = logic 1
c) If input A = logic 1	AND input B = logic 0	then output C = logic 1
d) If input A = logic 1	AND input B = logic 1	then output C = logic 0

Step 3

Next, just replace the NOT-AND title to NAND to give us:

NAND Gate Truth Table

Logic State	Logic State	Logic State
a) If input A = logic 0	AND input B = logic 0	then output = logic 1
b) If input A = logic 0	AND input B = logic 1	then output = logic 1
c) If input A = logic 1	AND input B = logic 0	then output = logic 1
d) If input A = logic 1	AND input B = logic 1	then output = logic 0

NOR (NOT-OR) Gate

The NOR gate is a logic device in which the output goes high (logic 1) only when both of the inputs are low; any other condition produces a logic low (0). For simplicity we are using here as an example a simple two-input logic gate. The NOR gate can be considered to be the combination of a NOT and an OR gate, that is, the truth table is the complement of the OR gate. For a two-input device, the truth table is as shown below. We can see quite clearly how the output behaves when stimulated with the indicated input combinations.

NOR Gate Truth Table

Logic State	Logic State	Logic State
1) input A = 0	input B = 0	output = 1
2) input A = 0	input B = 1	output = 0
3) input A = 1	input B = 0	output = 0
4) input A = 1	input B = 1	output = 0

The NOR gate can be considered to be the negation of the OR function. Once more, start with the OR function and follow the step sequence.

Step 1

Set up the OR gate truth table as shown.

Logic State	Logic State	Logic State
a) If input A = logic 0	OR input B = logic 0	then output = logic 0
b) If input A = logic 0	OR input B = logic 1	then output = logic 1
c) If input A = logic 1	OR input B = logic 0	then output = logic 1
d) If input A = logic 1	OR input B = logic 1	then output = logic 1

Step 2

Then negate it, by changing all the ORs to NOT-ORs and inverting the output states, so we get:

NOT-OR Gate Truth Table

Logic State	Logic State	Logic State
a) If input A = logic 0	NOT-OR input B = logic 0	then output = logic 1
b) If input A = logic 0	NOT-OR input B = logic 1	then output = logic 0
c) If input A = logic 1	NOT-OR input B = logic 0	then output = logic 0
d) If input A = logic 1	NOT-OR input B = logic 1	then output = logic 0

Step 3

Next, just replace the NOT-OR title to NOR to give us:

NOR Gate Truth Table

Logic State	Logic State	Logic State
a) If neither input A = logic 0	OR input B = logic 0	then output = logic 1
b) If neither input A = logic 0	OR input B = logic 1	then output = logic 0
c) If neither input A = logic 1	OR input B = logic 0	then output = logic 0
d) If neither input A = logic 1	OR input B = logic 1	then output = logic 0

Assume that at time T1 the two inputs are in a logic low (0) state. This is our starting state, and the output under these conditions is at a logic high (1). At the next time interval, T2, input B is taken to a logic high (1). The output changes to a logic low (0) state. At the next time interval, T3, the input states are swapped, with input A taken to a logic high (1) state and input B taken to a logic low (0) state. The output remains at the previous logic low (0) state. At time T4, both inputs are sitting at a logic high (1), and the output remains in a logic low (0) state. As we sequence the inputs down once more, the next time interval, T5, sees the input A at a logic high (1) and input B at a logic low (0). The output remains in a logic low (0) state. Finally, as we reach time T6, both inputs are again at a logic low (0). The output finally changes state to logic high (1).

NOR Gate Truth Event Sequence

Time	Input A	Input B	Output C
T1	0	0	1/starting value
T2	0	1	0/change of state
T3	1	0	0/no change of state
T4	1	1	0/no change of state
T5	1	0	0/no change of state
T6	0	0	1/change of state

EX OR (Exclusive OR) Gate

The EX OR is a logic gate device in which the output goes high (logic 1) when only one of the inputs is taken high (logic 1). When the inputs are the same, the output is in a low (0) state.

The EX OR gate is a special case of the generalized OR gate. Up to the first three logic conditions (1, 2, 3), both the EX OR and OR gate are the same. In the final logic condition, (4), the EX OR does not allow the output to switch to a high (1) state, that is, it is exclusively an OR device.

For a two-input device, the truth table is:

EX OR Truth Table

Logic State	Logic State	Logic State
1) input A = 0	input B = 0	output = 0
2) input A = 0	input B = 1	output = 1
3) input A = 1	input B = 0	output = 1
4) input A = 1	input B = 1	output = 0

Assume that at time T1 the two inputs are in a logic low (0) state. This is our starting state, and the output under these conditions is at a logic low (0). At the next time interval, T2, input B is taken to a logic high (1). The output changes to a logic high (1) state. At the next time interval, T3, the input states are swapped, with input A taken to a logic high (1) state and input B taken to a logic low (0) state. The output remains at a logic high (1) with no change in state. At time T4, both inputs are sitting at a logic high (1), and the output switches into a logic low (0) state. As we sequence the inputs down once more, the next time interval, T5, sees the input A at a logic high (1) and input B at a logic low (0). The output switches state once more to a logic high (1). Finally as we reach time T6, both inputs are again at a logic low (0). The output changes again to a logic low (0).

EX OR Gate Truth Table Event Sequence

Time	Input A	Input B	Output C
T1	0	0	0/starting value
T2	0	1	1/change of state
T3	1	0	1/no change of state
T4	1	1	0/change of state
T5	1	0	1/change of state
T6	0	0	0/change of state

EX NOR (Exclusive NOR) Gate

The EX NOR is a logic gate device in which the output will go low (0) only when the inputs are not the same. When all of the inputs are the same, the output is in a high (1) state. For a two-input device, the truth table is:

EX NOR Truth Table

Logic State	Logic State	Logic State
1) input A = 0	input B = 0	output = 1
2) input A = 0	input B = 1	output = 0
3) input A = 1	input B = 0	output = 0
4) input A = 1	input B = 1	output = 1

Assume that at time T1 the two inputs are in a logic low (0) state. This is our starting state, and the output under these conditions is at a logic high (1). At the next time interval, T2, input B is taken to a logic high (1). The output changes to a logic low (0) state. At the next time interval, T3, the input states are swapped, with input A taken to a logic high (1) state and input B taken to a logic low (0) state. The output remains at a logic low (0), with no change in state. At time T4, both inputs are sitting at a logic high (1), and the output switches into a logic high (1) state. As we sequence the inputs down once more, the next time interval, T5, sees the input A at a logic high (1) and input B at a logic low (0). The output switches state once more to a logic low (0). Finally, as we reach time T6, both inputs are again at a logic low (0). The output changes again to a logic high (1).

EX NOR Gate Truth Table Event Sequence

Time	Input A	Input B	Output C
T1	0	0	1/starting value
T2	0	1	0/change of state
T3	1	0	0/no change of state
T4	1	1	1/change of state
T5	1	0	0/change of state
T6	0	0	1/change of state

The table below presents a nice truth table summary of the six basic logic gate types.

Table 4-1 Truth Table Summary for Basic Logic Gate Types

AND			OR		
Input A	Input B	Output C	Input A	Input B	Output C
0	0	0	0	0	0
0	1	0	0	1	1
1	0	0	1	0	1
1	1	1	1	1	1

NAND			NOR		
Input A	Input B	Output C	Input A	Input B	Output C
0	0	1	0	0	1
0	1	1	0	1	0
1	0	1	1	0	0
1	1	0	1	1	0

EX OR			EX NOR		
Input A	Input B	Output C	Input A	Input B	Output C
0	0	0	0	0	1
0	1	1	0	1	0
1	0	1	1	0	0
1	1	0	1	1	1

CHAPTER **5**

Components Review and Projects 2–8

Electronic Components

A huge variety of components are available to you when starting a project build. The options can seem daunting, but when rationalized, the choice can be narrowed down quite easily to just a few basic components. There is nothing more fundamental than the resistor. Superficially it doesn't appear to perform any duties to shout about, but notwithstanding its deceptive simplicity, the resistor is one of the keys to all circuit designs. The other equally lowly (that is, when compared to the esoteric microprocessor) yet indispensable component is the capacitor. These two building blocks appear everywhere.

Given that you're interested in, and going to shortly get into, circuit builds, what follows is a nice introduction for the beginner, or a good refresher if you've already been experimenting with project constructions. Regardless of whether the circuit is digital (the focus of this book) or analog (my previous volume), there are certain sections, like the power supply, that are exactly the same (well, exactly the same as far as my own books are concerned). As the basic power supply (i.e., the type you'll see in all my circuit projects) features both our two building-block components, the resistor and capacitor, why not make a start there?

In keeping with my preference for simplicity and ease of design, all the circuits are powered by a single nine-volt battery, not only making them safe (no 110 V line voltage shock hazard to worry about) but also providing an automatic safeguard against excess current flowing through an inadvertent assembly connection error. You can only get so much current from a nine-volt battery (it's not a huge current supplier); the voltage will quickly start to fall even if the maximum short circuit current is flowing, and that in itself makes it intrinsically a current-limiting device.

The Basic Power Supply

A basic power supply consists of a battery, the on/off switch, a power-on indicator, and some way of smoothing transients generated by switching currents in circuits, especially digital gates. The power-on indicator is taken care of by the versatile light-emitting diode, or LED. A resistor takes care of the current-limiting requirement for the LED. An excess of current (if you had no current-limiting resistor in place) would swiftly curtail the LED's life. This is probably the most basic example of the need for a resistor! The illumination level from the LED is directly proportional to the current flow.

What's an optimum level? Certainly a minimum level is for the LED to be visible (when lit); on the other hand, it can't be so bright it dazzles (it's supposed to be a power-on indicator, not a beacon!). With our customary nine-volt supply, we're looking at resistor values that lie anywhere between 1 kohm and 10 kohm. So, often you'll see featured here a 4.7 kohm resistor for the LED current limiter. It's not a critical value, one that has to be adhered to; take it up or down (in value) and it'll still be acceptable, but don't forget it has to be there. A resistor's akin to the nuts and bolts holding together a piece of electronic equipment, like your latest audio system. Where would your megawatt system be without the hardware to hold it all together? Nowhere. We all have a role to play, no matter how humble.

The smoothing requirement mentioned earlier is taken care of by another seemingly lowly component, the capacitor. In this case (i.e., the power supply circuit we're looking at), it has one of the most common values you'll see figured in these projects, $0.1\,\mu F$. In the event that the nine-volt battery power is replaced by a nine-volt line adapter, you'll also see in my circuits a $100\,\mu F$ electrolytic capacitor shunting the smaller $0.1\,\mu F$ capacitor. This is so that any changes in the supply current generated as a result of circuit transients are smoothed out.

Now that you've seen the roles played by the resistor and capacitor, we'll go into more detail on individual components. Let's start with the resistor.

Resistors

A resistor is a two-terminal device having an electrical property of resistance. Resistance, as you've probably guessed already, is measured in units of ohms. Most typically in circuit designs, resistor values from ohms to kohms and perhaps a megohm will be encountered. The most popular value would be a 1 kohm value (my choice), based on most frequent appearance in a circuit. The resistor is an extremely robust component, able to take a lot of punishment (you don't have to handle it with care) and still come out on top. Short of taking a chainsaw to it, the resistor's practically indestructible. From an electrical viewpoint, there is much to like too.

Electrically, a circuit resistor presents a resistance to the current flow that would take place when a voltage is applied to a circuit. A high resistance presents more of an "obstacle," and so the current flow is relatively small. On the other hand, a low resistance allows more current to flow. If that were all there was to a resistor's function, that is, just controlling the current flow through a circuit, we wouldn't have much of a range of circuits to conjure with. But human ingenuity being what it is, electronics designers have come up with a lot more uses for the resistor. What can be done with, say, two resistors? Other than just limiting current, one of the most versatile functions for two resistors is as a potential-divider circuit. What happens when, say, two equal resistors are placed across a simple voltage source (e.g., a nine-volt battery)? At the junction of the two resistors, i.e., the midpoint, we would have half the supply voltage existing; in this case it would be 4.5 volts (as we're using a nine-volt supply).

This can be easily verified with a multimeter set to measure dc volts, placed across the midpoint and the negative battery voltage. We can determine the current flow in a resistor very simply, by dividing the applied voltage by the resistance. Our voltage source is (in this book) always going to be a nine-volt battery. So if we've got two 100 kohm resistors in series, the current is a minute 0.045 mA, as the current is the ratio of the voltage to the resistance. The potential divider, as the name implies, divides the voltage being applied to it. The two equal-value resistors are a special case, but the resistors could take on any other set of values. The output voltage calculation is given by the following simple equation:

$$\text{Output voltage} = \text{input voltage} \times R1/(R1+R2)$$

where R1 is the value of the resistor across which we're making the measurement and R1 + R2 is simply the sum of the two resistors.

Ohm's Law

Ohms's Law defines the relationship of the current flowing through a resistor when a voltage is applied to it by this simple relationship: voltage divided by resistance equals current flow. These measurements are all in the fundamental units of volts (for voltage), ohms (for resistance), and amps (for current). So taking our nine-volt source (our usual battery source) and a 4.7 kohm resistor (the LED current limiter), we can calculate the current flow as current = 9/4700 = 1.9 mA (milliamps). These are what I like to call "practical parameters." You can carry out so many simple experiments that are so easy to understand with just a battery, a handful of resistors, and a multimeter.

The Potentiometer

So, if we could somehow make the resistors (from the previous potential divider circuit) continually variable, could we not have a means of generating any voltage between zero volts and nine volts? Is there such a thing as a continuously variable resistor? Turn up your radio (if you've got it on). Behind that volume-control knob is a component called a potentiometer, or a continuously variable resistor.

In most applications where it's controlling volume, it's going to be wired up in a basic configuration like this: the incoming signal is applied across the outer terminals, and the output voltage taken from the center terminal and ground. The potentiometer has three terminals. A curved resistor track forms the main body of the component. By a clever piece of mechanical dexterity, a rotating arm is made to bear on the track under pressure, so that as the arm is rotated, the resistance either increases or decreases, depending upon which direction we're going. Feeding in the source voltage to the two outer terminals and taking the output from the center wiper terminal (and either of the outer two terminals) makes the output continuously variable as the shaft is rotated. If you had, say, a 100 kohm potentiometer fed with a nine-volt dc source (from a battery), the output can be varied from zero to the full battery voltage. Using a high-value potentiometer such as the 100 kohm limits the current drawn, and hence the battery voltage remains stable. A lower-value potentiometer, such as 1 kohm, could also be used, but the current drain is much higher (actually ten times more than with the 100 kohm potentiometer). The actual current values are 9 mA for the 1 kohm potentiometer, which is relatively significant, especially over time, but only 0.09 mA for the 100 kohm potentiometer, a really low value. The maximum resistance value will be that marked on the potentiometer, typically 1 kohm, 10 kohm, and 100 kohm.

Resistance Measurements

A multimeter set to the resistance range can easily check out resistor values. Bear in mind that while making resistance measurements, your body has a resistance too, so make sure your fingers do not touch both of the probes. Should this happen, you'll have your body's resistance in parallel across the resistor you're trying to measure, and that's going to give you a false reading. Of course, it (the degree of falseness) depends upon the body's actual resistance, which is a function of how moist your fingers are and the degree of pressure you're applying. Contact with one finger is fine, but it's better practice to avoid touching the leads in the first place. The other precaution is to zero-adjust the resistance meter first. You do this by shorting the meter terminals and adjusting the "zero" knob till the meter reads zero. This only applies to the analog type of multimeter.

Of course, to verify the resistance reading you're getting, you have to know how to read off the resistance value. Putting together a circuit design on paper is quite different from putting the circuit together with real components. That's where you're going to need some building smarts—and that's the purpose of this book (for digital circuits), not only to see how the basic digital circuit blocks are configured but also to take the concepts learned to actual working projects that can be experimented with, tested, and evaluated. Building up a knowledge base from the bare bones is how we're going to progress through this book.

So, how do we read off resistor values?

Resistor Color Code

The numerical value of a particular resistance is marked on the component body, typically as a three-color-band code for the actual resistance value. A fourth band represents the tolerance, but for simplicity this can be ignored if you just want to read off the resistor value (which is generally the case). As you almost certainly will want to look up the resistor color code, here it is.

Color Band	Equivalent Number Code
Black	0
Brown	1
Red	2
Orange	3
Yellow	4
Green	5
Blue	6
Violet	7
Gray	8
White	9

The most common resistor values in use range from 1 ohm to 1 Mohm. Although 1 ohm might be considered to be a short, it's not really. For most practical applications though, the 1 kohm value is usually the lowest value found.

With most circuit applications, you can cover the majority of circuit needs with just a few favorite values. My typical preference is: 100 ohm, 1 kohm, 4.7 kohm, 10 kohm, 100 kohm, and 1 Mohm. Take a look at the circuits in the book and see how often these values will figure in the parts list. Intermediate values can be built up easily by manipulating a handful of basic values and learning a bit of "resistor math."

Resistors in Series

The simplest way to build up resistor values is to add them together in a series mode. That means that one end of a resistor is physically connected to the end of the next one. Measurements are taken across the two free ends of the resistors, one from R1 and the other from R2. So, if we've got two resistors, R1 and R2, each 10 kohm in value, the series combination is just the sum R1 + R2, that is, 10 kohm plus 10 kohm, which equals 20 kohm. It's just simple arithmetic. Any value can be built up this way, but, of course, the physical size of the resistor chain increases with each addition.

The other way of coupling resistors is the parallel mode. Here the two resistors, R1 and R2, are connected with one end of R1 going to one end of R2, and the other end of R1 going to the other end of R2. Unless you've actually been shown this mode of connection, it's not obvious, especially if your talent lies more in the design side. In the special case of two equal-value resistors connected in parallel, the result is half the resistor value. So, two 10 kohm resistors produce a 5 kohm value, and two 1 Mohm resistors give you a 500 kohm value. If a circuit calls for an odd 5.5 kohm resistor, couple two 1 kohm resistors in parallel to produce 500 ohm, and then couple two 10 kohm resistors in parallel to give 5 kohm, add the two parallel networks, and there it is. Manipulating the various series/parallel combinations will produce almost any value you want.

There is a more general combining rule that takes into account the situation when the resistors are not equal in value. Here's the general rule: for two unequal value resistors parallel together, the total value is the product of the two resistors divided by the sum of the resistors. For example, if R1 is 10 ohm and R2 is 100 ohm, connected in parallel, the equation gives the resultant as the product, which is $10 \times 100 = 1000$ ohm, divided by the sum, which is $10 + 100 = 110$, i.e., $1000/110 = 9.9$ ohm. Another useful item to remember when connecting two resistors in parallel is that the total resistance is always less than the smaller of the two individual resistor values. In the example above, the resultant value of 9.9 ohm is less than 10 ohm (the smaller of the two resistors). It provides a quick check on the accuracy of your arithmetic.

What about the case of more than two resistors in parallel? What do we do then? Where we have more than two resistors in parallel (and you can have as many as you want), the rule is: 1/total resistance = 1/resistor 1 + 1/resistor 2 + 1/resistor 3, and so on. It's going to take a little care to do the calculation, as there's an extra step involved (as you'll soon see) toward the end of the calculation.

Look at an example of 10 ohm, 20 ohm, and 30 ohm resistors connected in parallel. We are going to get:

1/total resistance = $1/10 + 1/20 + 1/30 = 0.1 + 0.05 + 0.033 = 0.183$ ohm

As 1/total resistance is 0.1833, we do a simple inversion to get the total resistance, which is now $1/0.1833 = 5.455$ ohm. Checking this value, we can see that it (5.455 ohm) is less than the smallest resistor value, 10 ohm.

Resistor Tolerance

We've been dealing so far with actual resistor values. There is a second parameter associated with the resistance value. This is the tolerance parameter, which is represented by an extra color band on the resistor body. The most commonly specified tolerance is 5 percent (a gold band), followed by

10 percent (a silver band) tolerance. In case you see them, there are also resistors with no color band, which means a 20 percent tolerance. What is this tolerance? The tolerance percentage refers to the spread of values on either side of the nominally marked value (the three color bands) that the resistor is allowed to have. For example, if you have a 100 ohm, 5 percent resistor and measure the actual resistance, it could lie anywhere between 100 ohm + 5% = 100 ohm + 5 ohm = 105 ohm, or 100 ohm − 5% = 100 ohm − 5 kohm = 95 ohm. Tolerance is the spread of values you can expect the resistor to have. Obviously, we would want it to be as close to the marked value as possible. If we wanted a 100-ohm resistor, we expect to get that value and not 95 ohm. Not true! You're going to get what is marked on the body, within the stated tolerance bounds. The more sharply you want to limit the spread, the more costly the resistor is going to be. However, to give you a feel of what is acceptable, a 5 percent tolerance (as specified for the projects here in this book) is a good value to use. Were this a 20 percent resistor, the limits would run even farther from 120 ohm to 80 ohm—which is extraordinarily wide!

Power Rating

The third parameter associated in specifying a resistor is the power rating. Most typically the value used is 1/4 watt, which is also specified for the project circuits in this book. It's physically a nice, small size, fitting optimally into assembly boards and being not too weak. The resistor power rating defines the ability of the resistor to dissipate power when a current is flowing through the resistor. When current flows through a resistance, it gets hot! The more current you pass through a resistor, the hotter it gets, and hence the resistor must be able to stand up to the dissipated power. Larger resistors will have a rating of half a watt and even more. It's inefficient to specify a higher than necessary power rating, as they will just take up more board space and cost more.

Once more going through a little arithmetic, we'll see how the power rating is affected by series or parallel resistor combinations. Consider the simple case of two 10 ohm, 1/4-watt resistors connected in series. We would get a total resistance 20 ohm. The power rating stays at one-quarter watt. But what happens when these resistors are joined in parallel? Something interesting: the resistance drops to five ohms, as we know, but now the power rating is raised to half a watt, which is a valuable gain if your circuit has power to dissipate. Here's another example. Let's say you wanted a 10-ohm, one-watt resistor; this is quite a large device. You've got a lot of 100-ohm, 1/4-watt resistors on hand. Is there anything we can do? Indeed! Take ten of these 100-ohm resistors and parallel them all together. The total resistance, when you work out the math, is going to be 10 ohms (one tenth of the individual values), and the power will have increased to $10 \times 1/4 = 1.25$ watts. It's a useful tip to remember.

Capacitors

Next to resistors, capacitors are the most common component seen in circuit designs. Capacitors, though, are ac components. What does that mean? There are two ways circuits can be operated, either with dc or ac signals. A dc signal, such as a battery voltage, has a value that does not fluctuate over time. Display a battery voltage on an oscilloscope, and it'll be constant; it doesn't vary instantaneously over time. Display a sine wave on an oscilloscope, and it varies instantaneously periodically over time (the sine wave needs to be slowed down considerably in frequency in order to see this on the oscilloscope).

The battery is a dc signal; the sine wave is an ac signal. Resistors can be considered to be primarily responsive to dc signals, whereas capacitors can be similarly primarily responsive to ac signals. For the sake of completeness of definition, though, resistors are also responsive to ac signals, but generally this is a secondary consideration; capacitors don't pass dc signals, that is, they can be considered to be nonresponsive to dc signals. Capacitors as with resistors are two-terminal devices and exhibit the distinctive property of passing only ac signals while blocking dc signals. There are circuit situations where the ac signal has to be passed but the dc component blocked. One of these examples is where a power amplifier's signal is fed to a speaker via a capacitor. Radios, CD players, and TVs all have audio sections where the speaker is fed through a capacitor. Another area in which you'll always notice them is at the input and output section of ac amplifiers.

Units of Capacitance

Capacitors are measured in units of farads, but as this is a very large unit of capacitance, the much smaller units of pico, nano, and microfarads are most often used. A picofarad is 10^{-12} farads, a nanofarad is 10^{-9} farads, and a microfarad is 10^{-6} farads. The conversion between the units is that 1 pF equals $10^{-6}\,\mu F$.

Capacitor Marking Codes

There are three numbers (marked on the body of some capacitor) used to represent the capacitance value. For example, take a look at the common 0.1 µF capacitor that is used throughout the circuits in this book. There is a number marked on the capacitor such as "104." This will be the capacitance value expressed in picofarads. The first and second numbers define the first two digits of the capacitance. The third number defines the number of zeros to be added after the first two numbers. So "104" is equivalent to 100,000 picofarads. As this number's a bit unwieldy, multiply it by 10^{-6} to convert to µF, giving a value of 0.1 µF, which is a much more convenient number to work with. This is a very common capacitor value, as you'll later see when the projects are described. All of my circuits have a number of 0.1 µF capacitors liberally sprinkled.

As another example, what about a 470 pF capacitor? The capacitor is in the correct units of pF, first of all. If it's not, it has to be converted. The first two capacitance numbers are 4 and 7, so that's also the first two markings. There is one zero after the 47, so the number after the 47 is 1. Hence 470 pF has a marking of "471."

Unfortunately, unlike resistors, there is no easy way to check capacitor values. You don't know if they're OK, open circuit, or short circuit. But they're very reliable, so you just put them into a circuit and assume they're fine. It is usually the case. Resistors are so much more "appealing" from that point of view, as they're so easily checked with an ohmmeter.

Nonpolarized and Polarized Capacitors

The physical structure of capacitors basically falls into two specific categories. There's the simple nonpolarized type, which is physically small in size (such as the ceramic type) and small in electrical value (i.e., capacitance), and the physically larger polarized type, with higher associated capacitance values. The electrolytic capacitor is a polarized component and has markings on the body to indicate the appropriate negative and positive terminals, defining which way the device has to be connected in a circuit. Generally you will find that capacitors above and including 1 µF in value are polarized. Capacitance values for the larger component values are marked on the component's body, as there is sufficient space to print out the value "in full," that is, 1 µF will actually be printed as that. Smaller value capacitors use a unique numbering code (as explained previously) to represent the capacitance values, using a system similar to the resistor color code, except there are no colors.

Radial and Axial-Lead Capacitors

The mechanical structure of the larger electrolytics is further split into two formats. The capacitor leads can emerge from the same end of the body. This is called a radial-lead device. Where you don't have a height restriction, this is the type I would recommend, as it takes up less board space. It's the type used for the projects described here. The leads can also emerge, with one lead from either end of the body. This type is called an axial-lead device. It takes up a huge amount of board space and is only used when the height clearance is critical, which is not usually the case for simple projects (like those described here).

Series and Parallel Combinations of Capacitors

Capacitors can also be connected in series and parallel like resistors to form different values. But there is a major difference as far as the net effect goes. Recall that resistors add when connected in series. Well, with capacitors it's exactly the opposite! To increase a capacitor value, we parallel two capacitors together; when two 0.1 µF capacitors are connected in parallel, we get 0.2 µF. Three capacitors of 0.1 µF value each, connected in parallel, give us 0.3 µF, and so on. Now if the capacitors are connected in series, the total capacitance is given by 1/total capacitance = 1/capacitance 1 + 1/capacitance 2, and so on. This equation looks, of course, similar to that for resistors in parallel, so it's worked out in exactly the same way. For example, two 0.1 µF capacitors connected in series will result in a 0.05 µF capacitor, since 1/total capacitance = 1/0.1 µF + 1/0.1 µF = 10 + 10 = 20. Hence the capacitance = 1/20 = 0.05 µF. Sometimes for timing applications (using the LM 555 timer) you might want to change the output characteristics a bit, and this is one way of getting a 0.05 or 0.2 µF value if you don't have one handy.

Capacitor Working Voltage

Working voltage is the final parameter we need to know about capacitors. Depending on the supply voltage used, the value of the rated working voltage for the capacitor is related to that of the supply voltage. Where the supply voltage is 9 volts, a rating of 25 volts is a good value to use. It's twice the supply voltage, with an extra bit of headroom.

Variable Capacitor

Is there an equivalent to the variable resistor, otherwise known as the potentiometer? Variable capacitors do exist but are less common than variable resistors. Variable capacitors, often called trimmers or trimmer capacitors, though, are two-terminal devices. Why? To understand this we need to go back a step and see how a capacitor's capacitance is defined. A capacitor is always constructed with a sandwich structure. There are two metal or conducting plates, separated by an insulator or an air gap (which also is an insulator). Basically, the capacitance is a function of the area of the overlapping metal plates (the greater the overlap, the greater the capacitance) and of the separation between the plates (the smaller the separation, the greater the capacitance). The insulator, or dielectric, also increases the basic capacitance value, by a factor that is function of the insulator material. The variable capacitor has an air gap, with plates that are composed of one fixed and one variable. As you vary the overlap, by rotating the variable plate over the fixed plate, the capacitance changes, increasing as the overlap increases, and vice versa. Most trimmers are tiny devices, and the adjustment is affected by means of a tiny screw fixed to the movable plate. So you see, just two terminals are needed for a variable capacitor. There are no variable capacitors used in the projects here.

Switches

Switches occur in many places in circuit schematics. In spite of their somewhat mundane nature, there are many varieties. But isn't a switch just an "on/off" device? No, there are actually many different types, and it's a good idea to get to know the basis for the variations.

First of all, there are two descriptors specific to switches, "poles" and "throws." The simplest type, as you'd find for applying power, to a circuit is called a single-pole, single-throw, or SPST, switch. It has two terminals, the minimum number you could have with a toggle mechanism that flips back and forth. Switches always have to be described with respect to how an input signal is connected to an output signal. That is what switches do, just make or break an electrical connection in a circuit. The descriptor term "pole" refers to the number of terminals the input signal (for example, the nine-volt positive-voltage supply rail) can be connected to. When there is just one, we have a single-pole situation. The term "throw," on the other hand, refers to the number of terminals the output can be connected to. When there is just one, we have what is defined as a single-throw. If there had been two terminals available to which the output could be connected, there would be two throws, and this would be a single-pole, double throw (SPDT) switch. In the SPDT switch there are actually three terminals arranged in a row on the back of the switch. Connection from the input is always made to the center terminal, while the remaining two terminals are reserved for the two outputs. The SPST switch is used in the power supply line for the projects described here. The SPDT is a more useful switch, as it can double for the SPST and hence cuts down on the number of switches to keep in inventory. In cases where there is a need to have two sets of "single-pole, single-throw" actions occurring simultaneously, the SPST exists as a DPST (double-pole, single-throw) switch. The dual version of the SPDT is the DPDT (double pole, double throw), and it is more useful than the DPST. It minimizes inventory more.

Apart from the switching differences, switches are also available in various current-carrying capacities. Higher current capacity ratings usually indicate a physically larger switch.

There is no benefit to using a larger-current switch than necessary. In fact there's actually a practical disadvantage. Heavy-duty toggle switches having high current ratings require physically more force to toggle between the on and off positions. If most of your project builds are going to be housed in a small plastic case, that case is going to be very lightweight. A switch requiring an undue amount of force to toggle will very likely tip the case. This is an inconvenience. The more appropriate type of switch to use is the lightweight type, which requires the bare minimum of finger pressure to toggle. As switches are costly, it is a point worth remembering.

Jack Plugs and Sockets

Connections into and out of project cases are made much neater and easier with miniature 1/8-inch jack plug/jack socket combinations. It also gives it a much more professional look than wires trailing out. This 1/8-inch jack plug is almost always found on the end of headphones provided for portable radios and cassette players. The jack plug has a screw on barrel, often constructed from plastic, but sometimes a metal version can be found too.

There're two types available, the so-called mono and stereo types. These are not interchangeable, so take care that the socket matches the plug—although the stereo type can be wired as a sort of mono type, with both the signal channels wired together. Once the jack plug cover is removed, you'll see two connections, if it's a mono type. There's a short connection to the center pin and a longer connection that goes to the ground terminal. A mono jack plug can be recognized by the fact that it has a single narrow insulator strip located near the end of the jack plug tip. The stereo jack plug has two such insulator strips. Jack sockets come in the "normally closed" and "normally open" types.

For a basic application, such as connecting a speaker to an amplifier output, it makes no difference which type is used; we generally use the "normally open" type here. The "normally closed" type of socket has a special use, though. For example, where an amplifier is connected normally to an internal speaker, as you'd find in a regular transistor radio, this is the type of socket used. When an external pair of headphones is plugged in, the internal speaker is disconnected by the action of this jack socket.

Project Start

With that brief introduction to basic components, we can make a start into the project circuits, as this is the ideal way of gaining practical familiarity with components. The emphasis of this book is digital, of course. The integrated circuits highlighted are digital, belonging to the well known TTL family. There is one exception, though. We feature one linear device—that is, an analog device—but it's a rather special one. This is the popular LM 555 timer. It's popular because:

1. It's an extremely easy to use and versatile device that is found in the vast majority of circuits.
2. It is the simplest device available for generating extremely versatile waveforms.
3. It is widely available from any hobbyist component store.
4. The component values are noncritical.
5. The number of components needed to generate a convenient pulse train is small.
6. The output is compatible with TTL.

7. It will run off the same five-volt TTL supply voltage.
8. It is very stable in operation.
9. It is nontemperamental, that is, it works always when you switch on.
10. It is low cost.

There is no equivalent TTL generator that will offer the same versatility. If there were, it would have been placed here instead. Although the LM 555 can handle a supply voltage of 18 volts, the output will rise too and be non-compatible with TTL. By itself, the LM 555 will conveniently operate off a nine-volt battery. But to make it TTL compatible, the nine-volt supply should be fed through a five-volt regulator; that'll ensure the LM 555 output is TTL compatible. For a stand-alone generator, it's a good idea to mark the outside of the project case "TTL compatible output," so there's no chance of accidentally damaging any TTL circuit it feeds into. The first project build to be described therefore will feature the versatile LM 555 in a digital application.

Project 2: LF TTL Pulse Generator

Introduction

Undoubtedly the single item of equipment that you're going to be using the most will be a pulse generator, to act as a stimulus for testing your digital project builds. A signal source is essential to verify correct circuit operation. Digital circuits are essentially pulse-state changers, so we want to see the pulse change state correctly. That's what this project does for you. Using a handful of readily available and noncritical components, we're going to put together a pulse generator with a handy two-range, high/low frequency switchable capability and a potentiometer-driven frequency control for fine control. The low frequency range produces a sufficiently slow train of pulses to allow pulse transitions to be easily observed by eye, so you can make out the changing pulse states.

Unlike the customary output-signal amplitude control seen on most other generator circuits, as this project is a TTL-specific design, no such control is included. Why? Because this test generator is going to be used exclusively for driving and testing digital logic devices (the topic of this book), the pulse output must always be compatible with a TTL circuit (running from a five-volt supply voltage). If we had included an output control, it could inadvertently be user-set to the incorrect value (if there's a control provided, chances are it's going to be varied) and hence not provide the correct stimulus level. This way we remove any reason for the circuit under test to malfunction and waste time in needless troubleshooting.

The low frequency of operation of this unit (which is different from most other generator designs focusing on high speed) justifies the elimination of special shielded cable (coupling out the output signal) or BNC (high-frequency) connectors. A standard 1/8-inch jack plug and socket combination (the type found on portable stereos and CD players) is all that is needed. It's not sophisticated, but it does the job very well, and that's the principal requirement. Similarly, the recommended project case is a regular, easy to use (for drilling) plastic item, readily available from hobbyist stores.

We're not complete yet! There's still the monitor innovation. Stimulation is effected with the pulse generator. What is happening to the circuit under test? How do you know it's changing correctly? The simple adder (the LED monitor) included in this design does that for you! It's just an LED and series current-limiting resistor that's terminated in another miniature 1/8-inch jack socket on the front panel of the pulse generator. Couple in a matching 1/8-inch jack plug terminated with two miniature alligator clips, and you're in business. It's not high tech, but it's incredibly useful, and (to me) that's what counts. Time and time again you're going to be faced with the need for a quick digital verifier, and rather than wading through tangles of cables, here it is, conveniently located on the front panel. Oh yes, it also doubles as a useful power supply voltage. If it's on when coupled across the live and ground lines, you've got power!

But we're not finished yet on innovation. There is also an onboard LED monitor. This one verifies the pulse generator. Because the LED draws current, there's a switch to take it out of circuit when not required. Notice that the LED is brighter, because we've got a low-value resistor (1 kohm), as you want to see this LED light up clearly.

Circuit Description

The LM 555 (IC1) is an eight-pin IC in a dual-in-line package plugging into a mating socket on the assembly board. The schematic is shown in Figure 5-1. The use of IC sockets is well worth the minimal extra cost and effort, as it gives you the freedom to replace ICs (if necessary) without desoldering pins (a really time-consuming chore).

In the LM 555's free-running, or astable, mode, a square-wave generator is easily constructed from just a handful of components. The critical timing components are just two resistors and a capacitor. Although the duty cycle (ratio of the on period to the total on + off period) varies when the frequency is varied, this is not a serious issue and is a very small price to pay for ease of design. A variable resistor, VR1, provides a continuous adjustment of frequency over a specific range that is determined by the selection of either of the timing capacitors, C1a (1 µF) or C1b (0.1 µF). The higher the capacitor value for C1, the lower will be the frequency/frequency range you'll get, and vice versa. Switch S1 is a simple single-pole, double-throw type that does the selection for the timing capacitor (C1). Fixed resistors R1 (1 kohm) and R2 (10 kohm) also form part of the timing chain. Resistor R1 is the component connected from pin #7 to the positive Vcc supply voltage. Resistor R2 is added to prevent the resistance of potentiometer VR1(100 kohm) from going to zero at one end of the wiper travel. R2 is connected between pin #7 and one end of VR1. The other end of VR1 (actually the wiper terminal) connects to pin #6.

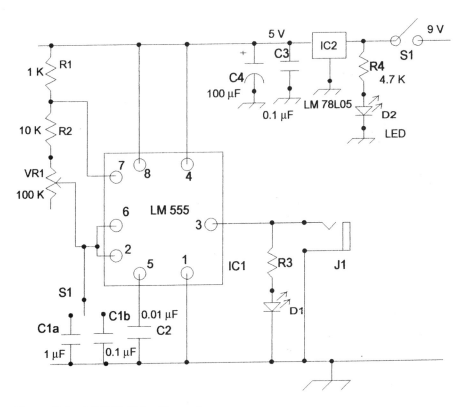

Figure 5-1 LF TTL Pulse Generator

As in all LM 555 free-running modes, pin #6 and pin #2 are connected together. A timing capacitor would normally run from pin #2 to ground, but as we have a selector switch included, the connection is a little more involved. The center terminal of the SPDT switch, S1, is routed to pin #2.

From there the two capacitors C1a and C1b are connected to the two outer switch terminals. Both of these capacitors terminate to the ground rail. Decoupling capacitor C2 (0.01 µF) is used merely to support the oscillator in the correct operating mode and is coupled between pin #5 and ground. The balance of the connections to IC1 are from pin #1 to ground and pin #8 and pin #4 to Vcc. The output waveform is a square wave taken from pin #3 routed through a miniature jack socket (J1) for convenience in monitoring the output waveform. There's a monitor LED (D1) and current-limiting resistor, R3 (4.7 kohm) wired across the output. Notice that for this project the square-wave output is taken directly from pin #3 without going through the usual coupling capacitor you'd find in an audio application. Why? This is because we need a TTL-compatible pulse, that is, one that makes a transition from zero volts to close to five volts. If the usual output capacitor were added, the resultant voltage would be incorrect for TTL compatibility. It would be positioned on the zero-voltage rail (since the dc component has been removed) and fluctuate equally above and below that value, somewhere close to half the supply voltage in the positive direction and half the supply voltage in the negative direction.

Using the 1 µF (C1a) timing capacitor (with switch S1 in the "a" position), we can span a frequency range from 6.5 Hz to 68 Hz, when potentiometer VR1 is rotated from a maximum to a minimum value. The frequency is inversely proportional to either the timing resistors or the timing capacitor. The component values for the timing components are chosen because these are the most commonly available and used values. If you wish, you can try lowering the frequency with a capacitor value that is doubled in value from 1 µF to 2 µF. This is easily done by just adding another 1 µF capacitor in parallel (which raises the value) across C1. The low-end frequency of 6.5 Hz would be dropped to 3.25 Hz. That's how simple and easy it is; it's a nice, quick way of tailoring the frequency to the value you want. If the logic states change too quickly to see by eye, just lower it to suit yourself. Generally though, the values given in the parts list will be the best starter values. In the "b" position of the switch S1, the lower-value capacitor of 0.1 µF (C1b) is switched, and now the frequency range scoots up from 65 Hz to 68.5 Hz. As the capacitor ratio for C1a and Cb is ten to one, the frequency range varies by the same ratio. The smaller the capacitor, the higher the frequency. The frequency should increase as VR1 is rotated clockwise. If this is not the case, reverse the connections to VR1. Toggle switch S1 to verify the change in the frequency band.

Looking at our schematic, we see that (conventionally) the power rail is drawn at the top of the schematic and the applied voltage is drawn on the right-hand side. The next component in from the battery is always the power-on/off switch, S2. To the left of the switch is the LED (D2) and its associated resistor, R4 (4.7 kohm). The resistor limits the current through the LED, and the value for R4 is a compromise between saving current and being able to see the LED; around 4.7 kohm is a good starter value. Feel free to vary it, and if you've got another close value, use that instead. The LED obviously is used as a power-on/off status indicator, and that's a very useful function, as the battery could rapidly drain if the power were left inadvertently on. Even at a few mA of current flow, a nine-volt battery doesn't last long.

Two capacitors decouple the power supply to ground. These are located close to the LED on the schematic, but physically on the assembly board, of course, they could be quite a distance away; it's the electrical connection that counts. These two components smooth out the power supply and enhance the stability of the supply line. The values are commonly 0.1 µF for C3 and 100 µF (an electrolytic type) for C4.

You'll begin to get familiar with what typically used values are for certain components like these after going through several of the projects. After a while you'll be able to define these component values almost without thinking. That's the philosophy behind the way these circuits are designed and presented here. Apart from the introduction to different digital circuits, it's also a way to learn how to put a circuit together from scratch. Power is supplied from a nine-volt battery, but with a difference for just this circuit. The nine-volt is stepped down to a stable five-volt value through the use of the three terminal voltage IC2 (LM78L05).

Formula for Frequency Calculation

The calculation for the frequency of oscillation is given by:

$$f(Hz) = 1.44/[R1 + (2 \times R2)] \times C$$

where

f (in hertz) is output frequency
R1 (in ohms) is the resistor going from pin #7 to Vcc
R2 (in ohms) is the total resistor value between pin #7 and pin #6
C (in farads) is the capacitor between pin #6 and ground.

Circuit components
R1 = 1 kohm
R2 + VR1 = 110 kohm
C1 = 1 µF

Output frequency
$f = 1.44/[1000 + (2 \times 110,000)] \times 1.10^{-6}$
$= 1.44/[1000 + 220,000] \times 1.10^{-6}$
$= 1.44/[221,000] \times 1.10^{-6}$
$= (1.44.10^6)/221,000$
$= 6.5 \, Hz$

Circuit components
R1 = 1 kohm
R2 + VR1 = 10 kohm
C1 = 1 µF

Output frequency
$f = 1.44/[1000 + (2 \times 10,000)] \times 1.10^{-6}$
$= 1.44/[1000 + 20,000)] \times 1.10^{-6}$
$= 1.44/[21,000] \times 1.10^{-6}$
$= (1.44.10^6)/21,000$
$= 68.5 \, Hz$

Circuit components
R1 = 1 kohm
R2 + VR1 = 110 kohm
C1 = 2 µF

Output frequency
$f = 1.44/[1000 + (2 \times 110{,}000)] \times 2.10^{-6}$
$= 1.44/[1000 + 220{,}000] \times 2.10^{-6}$
$= 1.44/[221{,}000] \times 2.10^{-6}$
$= (1.44.10^{6})/442{,}000$
$= 3.25\,\text{Hz}$

Construction Tips

Phase 1

As a rough check of the functioning of the basic oscillator circuit, use a single capacitor for C1 (leave out the switch for the time being). Potentiometer VR1 can also be left out until later on, and the connection from pin #2 (of IC1) can be taken directly to R2. All the power-supply-associated components located at the top of the power supply line can be left out for now, too. With all this in mind, the circuit becomes very simple, and with so few components, the chances of getting the circuit working the first time are very high. Add the LED and resistor limiter across the output from pin #3. When you switch on the power, this monitor should flash in sync with the high-state transitions.

Phase 2

Now you add the rest of the components as per the schematic, knowing that the basic circuit works. This is a philosophy I always use when building a new circuit: this partial build-test philosophy improves the probability of an error-free build. It's so easy to make a connection to the wrong spot (the usual reason for circuit errors), and the more connections there are, the more the chance for error. Socketing the design for the IC allows for very easy replacements (hardly likely, but just in case). Without an IC socket, removal of an IC requires a lot of dexterity, with a high risk of the board being damaged in the process. Finally, add the jack socket (J1) to the front panel. You should predrill the holes to fit the LED and jack socket. The supply monitor LED (D1), resistor R4 (1 kohm), and SPST switch (S1) complete the final build.

Test Setup

Testing should present no problems, as a result of our build-test-build strategy. If you have access to an oscilloscope, here's what you should check for, with the circuit running.

The amplitude of the output waveform sits at the zero volts level and makes a transition to a positive voltage level close to the five-volt supply value.

The frequency is increased when the smaller value of C1b is switched in, and vice versa.

The frequency is increased as potentiometer VR1 is rotated clockwise.

The output high condition should be in sync with the LED flashing on.

Parts List

Semiconductors
- IC1: LM 555 timer
- IC2: LM78L05 five-volt regulator

Resistors (All resistors are 5 percent 1/4 W)
- R1: 1 kohm
- R2: 10 kohm
- R3: 4.7 kohm
- R4: 4.7 kohm

Capacitors
(All nonpolarized capacitors disc ceramic)
(All electrolytic capacitors have a 25 V rating)
- C1a: 1 μF
- C1b: 0.1 μF
- C2: 0.01 μF
- C3: 0.1 μF
- C4: 100 μf

Additional materials
- VR1: 100 kohm potentiometer
- D1: LED
- D2: LED
- S1: single-pole, double-throw miniature switch
- S2: single-pole, single-throw miniature switch
- J1: 1/8-inch miniature jack socket
- Plastic project case
- Power supply: nine-volt battery

Project 3: High-Low State Display

Introduction

If you don't have access to an oscilloscope, here's a nice circuit demonstrating how two LEDs can be used as separate monitors for the on and off periods of a square wave. There's a LM 555 timer–based circuit for use as a reference generator. You might not have a suitable source circuit available, and in any case, this one's designed specially to showcase the on-period/off-period monitoring circuit. Circuit innovation is not the exclusive domain of design complexity. Simplicity pays dividends too.

Circuit Description

IC1 is the LM 555 timer configured in the astable operating mode in order to produce a very slow train of pulses. The schematic is shown in Figure 5-2.

These pulses have a pulse-repetition rate, or frequency, of 0.3 Hz. This low frequency has been specifically selected in order to demonstrate the function of the dual LED monitors. The frequency is controlled by three timing components: resistor R1 (10 kohm) connected between pin #7 and Vcc, resistor R2 (100 kohm) running between pin #6 and pin #2; and capacitor C1 (22 µF), which goes from pin #2 to ground. Note that pin #2 and pin #6 are shorted together when the LM 555 is configured in the free-running mode. Capacitor C1 is a large-value electrolytic, with the positive terminal going to pin #2 and the negative end returning to ground. The larger this capacitor's value, the lower will be the output frequency of the generated waveform. From pin #5 a small capacitor C2 of value 0.01 µF supports the necessary operating condition for the IC. It's a component that's always there and needs to be added to the circuit design as a matter of course. The three remaining connections complete the power and ground requirements and are similarly mandatory; pin #1 goes to ground, and pins #4 and #8 go to Vcc. The output waveform is taken from pin #3. Even though the output is a train of continuous pulses, it can be considered to be a dc-based signal, since it is positioned on the zero

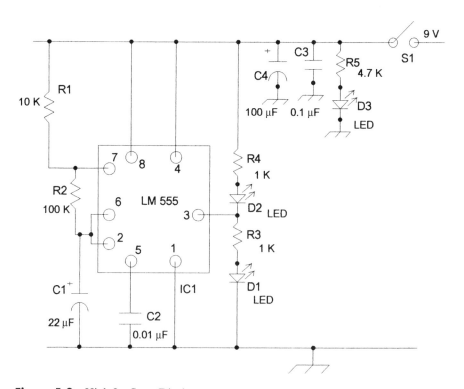

Figure 5-2 High-Lo State Display

voltage level and makes an excursion to the positive Vcc level during the "on" state of the waveform. Because of that, the simple monitor circuit can be configured.

Let's look at the lower half of the dual monitor circuit, which is the more conventional circuit. This is provided for by the light emitting diode LED1 (D1), and its current-limiting resistor R3 (1 kohm). From the way the LED1 (D1) is connected, we know that it'll turn on only when it is forward biased, i.e., when a positive voltage is applied to the anode end of D1. With each time the pulse appears, LED D1 will light up. Because of the slow pulse speed, you could gauge the number of pulses occurring per second and get an approximate idea of the frequency. Now look at the second LED (D2) monitor. The anode end of D1 is connected to the positive Vcc supply, so we know that LED D2 will conduct only if the cathode end is taken to ground. As that end of D2 is coupled through its current-limiting resistor, R4 (1 kohm), to output pin #3, and we know that in the "off" state pin #3 sits at the zero voltage level, the LED (D2) will turn on during the timer's off state—and it does.

In the free-running mode, therefore, as IC1 cycles through its on/off state, you'll see D1 and D2 turn sequentially on. Relative time durations between the on and off states can be very easily determined, more so than if we conventionally have just the lower LED (D1) in place. It would be less informative about the pulse's off state if that were the case. You could verify this by just disconnecting or covering up LED D2. Power for the circuit is supplied from a nine-volt battery feeding into switch S1. LED (D3) and current-limiting resistor R5 (4.7 kohm) tells you when there's actually power applied to the circuit. Capacitors C3 (0.1 µF) and C4 (100 µF) provide a smoothing function in the supply voltage line.

Here's the equation governing the frequency produced from the source circuit.

$$f(Hz) = 1.44/[R1 + (2 \times R2)] \times C$$

Circuit components
R1 = 10 kohm
R2 = 100 kohm
C1 = 22 µF

Output frequency
$f = 1.44/[10,000 + (2 \times 100,000)] \times 22.10^{-6}$
$= 1.44/[10,000 + 200,000] \times 22.10^{-6}$
$= 1.44/[210,000] \times 22.10^{-6}$
$= (1.44.10^6)/4.62.10^6$
$= 0.3 \, Hz$

Construction Tips

For the circuit build, it's advisable to build and keep this circuit as a reference device; that way, you've always got a nice control circuit. To simplify construction, it's preferable to have some sort of positive common rail running across the top of the assembly board. Similarly, across the bottom you could run a common ground rail. Start the build by putting the basic components such as the IC socket down first, followed by the wire links for the ground and Vcc connections. Don't forget the shorting link between pin #2 and pin #6. After that, start with the resistor R1, going to Vcc. Start with the resistors R1 and R2. After that add the capacitor C1, noting the polarity; the positive end of C1 goes to pin #2. The power line components can be left until later. From output pin #3 connect resistor R3. The free end of R3 terminates to a floating solder point. From there connect the anode end of LED D1. The cathode end of D1 terminates to the ground rail. Similarly, add the second resistor R4 to pin #3, with its free end going to a floating terminal. From that end make a connection of LED D2's cathode. Finally, the anode of D2 goes to the Vcc positive rail. The power section starts in from the nine-volt supply end. The positive supply voltage goes to the switch S1. The other terminal of S1 provides the feed voltage for the circuit. As this is a demonstration circuit to showcase the dual monitors, we don't have the five-volt regulator included here. The power-on LED (D3) feeds off from this point. Resistor R5 (4.7 kohm) provides current limiting. The usual two supply smoothing capacitors C3 (0.1 µF) and C4 (100 µF) follow next. Apart from the power switch S1, the power related components (D3, R5, C3, and C4) can be mounted directly on the circuit board. Use a nine-volt battery snap to couple power into the circuit.

Test Setup

There's no external connection needed to test this circuit as everything is already provided for on board. Just switch on and check out the monitor LED's D1 and D2. D1 shows the on period duration and D2 shows the off period duration.

Parts List

Semiconductor
    IC1: LM 555 timer
Resistors (all resistors are 5 percent 1/4 W)
    R1: 10 kohm
    R2: 100 kohm
    R3: 1 kohm
    R4: 1 kohm
    R5: 4.7 kohm

Capacitors
(All nonpolarized capacitors disc ceramic)
(All electrolytic capacitors have a 25 V rating)
    C1: 22 μF
    C2: 0.01 μF
    C3: 0.1 μF
    C4: 100 μF
Additional materials
    D1: LED
    D2: LED
    D3: LED
    S1: single-pole, single-throw miniature switch
    Power supply: nine-volt battery

Project 4: Pulse Speed Reducer

Introduction

Up to now the emphasis has been on producing a slow enough pulse frequency in order for the logic state changes to be viewed by an LED. But in real circuits we are more often than not going to be limited by the circuit components and not be able to slow down the frequency in order to get a visible display. There is another way of doing this, and provided the source frequency is not too high, we can get a slowed down pulse wave form that can be detectable visually. This is the pulse speed reducer circuit shown here. The heart of this circuit focuses on what is called a frequency divider. If we could produce one output pulse for every, say, 10 pulses coming in and display this series of single pulses we would have a pulse speed reducer. That's the principle of the circuit here. It's a very simple circuit to construct and since there are so few components needed it could very easily be built into a nice compact little unit that would serve nicely as a portable pulse tester.

Circuit Description

A single IC1 (CD4017) is the heart of our cool little circuit shown in Figure 5-3. The CD4017 is a CMOS decade counter device that runs off a nine-volt supply making it ideal for interfacing to the LM 555 timer circuit. There are also TTL counters available but these are limited to running off a five-volt supply. The CD4017 counter instead will run off a very wide supply voltage of 3 to 15 volts. So it will interface directly to a LM 555 running from a nine-volt supply or to a digital TTL circuit running off five volts. The LM 555 timer incidentally will run from 4.5 to 18 volts supply. These wide supply voltage ranges compare very favorably to the narrow five volts supply restriction of TTL. Other than the IC itself there are no other components needed to get this

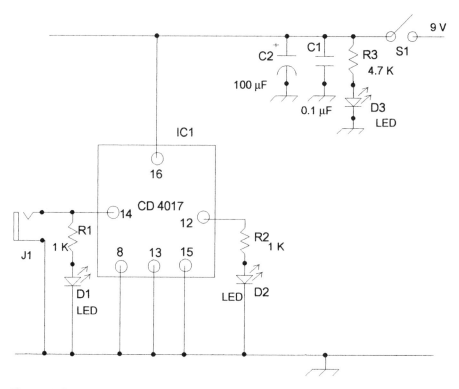

Figure 5-3 Pulse-Speed Reducer

circuit up an running. It has to take the record for the minimum component set for a circuit!

There's an input LED (D1) and current limiter resistor R1 (1 kohm) to give a visual indication of the feed signal. The output monitor, is another LED (D2) with current-limiting resistor R2 (1 kohm). Pin #8 is the ground connection. Vcc is taken from pin #16. IC1 is a 16-lead DIL (dual in line) package, so the IC socket is somewhat larger than our customary eight-lead device (e.g., for the LM 555 timer). Two further unused pins need to be grounded also for correct operation. These are pin #13 (the clock enable pin) and pin #15 (the reset pin). It's not necessary to know what these functions are for this pulse speed reducer circuit. The input signal is dc coupled to pin #14 defined as the clock input. The output emerges from the pin #12. That's all that's needed. The CD4017 is actually a versatile decade counter but the output pins are not used here. For reference purposes these output pins are #1, 2, 3, 4, 5, 6, 7, 9, 10, and 11. This circuit automatically divides down any input frequency by a factor of 10, e.g., so a 1 kHz input would emerge as a 100 Hz signal. Relatively high-speed signals will essentially cause the LED monitor to appear continually on. It's actually pulsing but the eye cannot discern the on/off transitions. You can also cascade this circuit by duplicating it and feeding the output from one into the input of the second. This total effect will be a 100 times slowing down of the originating pulse speed.

The usual power supply components can be found at the power supply input. The nine-volt battery is fed to on/off switch, S1. The feed from S1 drives an LED (D3) and current limiter resistor R3 (4.7 kohm). The two usual capacitors C1 (0.1 µF) and C2 (100 µF) smooth the supply voltage.

Construction Tips

The CD4017 is a CMOS device so static handling precautions must be adhered to, otherwise there's potential for damage to the IC. Modern CMOS devices can be destroyed by incorrect handling as a result of static charge buildup. This can result simply by just shuffling your feet or walking across the room. Integrated circuits are especially susceptible to damage that result when a source of electrostatic charge is discharged through the device. This is done just by touching the device. The degree of susceptibility to electrostatic charges varies depending on whether the device is bipolar and CMOS. Integrated circuits such as the LM 741, LM 555, and LM 386 are examples of bipolar devices. CMOS circuits are particularly sensitive to damage from electrostatic charges and the container packages housing such devices are always marked as such when you purchase these devices. These labels carry warnings, informing the user to take care when handling these devices. Bipolar devices are less sensitive and do not carry such warning labels. Digital logic such as TTL also belong to the bipolar family, though not all digital should be generically classified so—there also exists CMOS digital, just to confuse the issue! Integrated circuits will always be marked with an "ESD-sensitive" warning label if applicable, so you don't have to know whether it's CMOS or otherwise.

Test Setup

There's very little to go wrong with this circuit, but you're going to need a square-wave signal source to feed into the input terminal. The actual frequency range of your generator source is not critical as the intent is to demonstrate the speed reducing capability of this circuit. Somewhere though in the low kHz region would be a good start. It is important though to have a compatible signal level. So there are two options you can take:

1. Couple in a LM 555-based pulse generator and have both circuits driven from a nine-volt supply, *or*
2. Use a pulse source that is TTL compatible, that is, one that produces a zero- to five-volt output signal and also drives the pulse reducer circuit from a five-volt supply. This is easily done by using a five-volt regulator IC to get the nine-volt reduced down to five volts. As an example, the earlier circuit using the LM 555 timer has the five-volt regulator.

Parts List

Semiconductor
    IC1: CD4017 decade counter
Resistors (all resistors are 5 percent 1/4 W)
    R1: 1 kohm
    R2: 1 kohm
    R3: 4.7 kohm
Capacitors
(All nonpolarized capacitors disc ceramic)
(All electrolytic capacitors have a 25 V rating)
    C1: 0.1 µF
    C2: 100 µF
Additional materials
    D1: LED
    D2: LED
    D3: LED
    S1: single-pole, single-throw miniature switch
    Power supply: nine-volt battery

Project 5: Threshold Level Detector

Introduction

This circuit is unusual in that it is part analog and part digital. Since it has a digital component it's included here. Another way of looking at it is to examine the circuit's coupling arrangement. A general definition that I use is, a purely analog circuit will have the customary input and output coupling capacitors, as you would find in a conventional audio preamplifier circuit. Circuits, under this definition, that don't have these coupling capacitors are digital. Looking at this circuit you'll see no such capacitors; hence it's "digital." A slowly changing dc voltage can be found in circuits where a conventional audio signal has been demodulated; that is, it has been rectified and smoothed out to produce a dc voltage that is proportional to the amplitude of the incoming ac signal. The incoming ac signal is difficult to monitor, unless you have an oscilloscope as a monitor. By doing the demodulation, you get a slow-moving dc voltage that can be displayed on a regular analog dc voltmeter. On that basis, this circuit allows to you set a threshold level that, if exceeded, will trigger a monitor LED. This circuit thus shows you when a certain level has been exceeded—a convenient circuit to have, if you want to be sure of limiting your signal within a ceiling. This circuit could be used, for example, if you didn't want to drive an audio amplifier (accepting the original ac signal) into distortion. You would include this threshold-level detector as a parallel monitor. This circuit is useful for monitoring signal overloads.

Circuit Description

The heart of the circuit, IC1 (LM 324), is based on a simple analog device, an ordinary operational amplifier here, configured to operate in a dc mode. The schematic is shown in Figure 5-4.

The configuration of an operational amplifier can be determined by looking for the coupling capacitors at the input and output; if these are not here, you know that the circuit has been designed or configured to be a dc amplifier. The difference is critically important, since the circuit types (i.e., dc or ac operational amplifiers) are not interchangeable. Any ordinary operational amplifier, such as the LM 741, can be used as IC1. However, to be a little different I've chosen to use the LM 324 IC. This is a quad amplifier package, that is, the package contains four separate operational amplifiers, but only one is being used here. ICs of this type are so low cost and so easy to use that it's of little consequence that the rest of the three operational amplifiers are spare.

The LM 324 is quite different from the more familiar workhorse LM 741; the package size is different, and so too are the power and ground connections. If you look around other circuit projects you'll see the LM 324 showing up

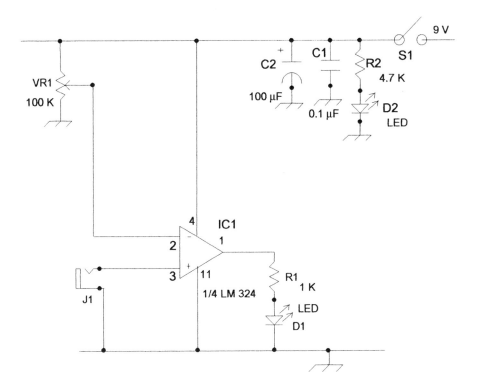

Figure 5-4 Threshold-Level Detector

quite a lot too, especially in ac amplifier situations. The LM 324 is a 14-lead DIL device with a somewhat unusual power and ground pin location. Generally, in such integrated circuits as the LM 741, LM 386, and LM 555, you'd find the power pin at the top half of the IC. For the purpose of orientation, the IC is positioned so pin #1 is located at the lower left-hand corner. This is the usual way the IC is laid out on any circuit board. The upper half of the IC includes the last pin. With the LM 324, the power pin is located on the lower side (the side containing pin #1) of the package, and the ground pin is along the top half. This reverse convention can be a source of wiring error if you're used to having the power pin sitting on the top half. Take special care, as ICs can be damaged if the supply voltage polarity is incorrect. Why have we chosen to use the LM 324 over the LM 741? Well, the circuit described here is a nice, easy one, and it is ideally suited to introducing you to a new IC—one that is very often seen in multi-amplifier based projects, which can seem a little daunting when all four amplifiers are used. So, here a simple one-quarter of the LM 324 is used, and we can thus become gently acquainted with it.

Returning to our circuit, we see that power goes to pin #11, and ground is routed to pin #4. The dc input signal goes to the noninverting pin (pin #3) via jack socket, J1, and the output emerges from pin #1. Remember that operational amplifiers always have a noninverting and inverting input terminal. When the noninverting pin #3 is used as the input feed terminal, it means that the output will change in phase with the input signal; if the input increases in amplitude, the output amplitude will increase also. It's too confusing otherwise to have the signals changing in antiphase if the inverting pin were used otherwise—hence the reason for choosing this configuration. When we're designing ac amplifiers, it really doesn't matter which signal pin is used. With our circuit here, the inverting terminal (pin #2) has an unusual connection. There's a potentiometer, VR1 (100 kohm), wired as a potential divider across the power and ground rails, with it's center terminal connected to pin #2. This part of the circuit will provide the dc reference voltage for our threshold detector. The potentiometer (VR1) allows us to vary the reference voltage.

Assume that a slow-moving dc input signal is applied to the noninverting terminal. When that signal exceeds the reference level, the output will trip and change logic state from logic low (zero voltage) into a logic high (Vcc) state. Neat! If we assumed also for the moment that the input signal were a sine wave that cycled smoothly through its waveform, the output would follow in step and result in a square wave being generated. This would give us a nice sine-wave to square-wave converter—another interesting way of looking at this threshold circuit! Unfortunately, it's not quite as easy going the other way (trying to get a sine wave from a square wave).

This circuit is technically what we call a comparator, as we are comparing the level (i.e., amplitude) of an input signal against the level of a reference signal. It's interesting to see the various ways we can title the exact same circuit! At the output end of the threshold level detector (taken from pin #1), we have a regular LED (D1) and limiter resistor R1 (1 kohm). The

LED (D1) will light up every time the reference level threshold is exceeded, and soon as the input signal level drops again below that threshold, the output will go low again (where it normally sits in the absence of a signal), extinguishing D1. That's all there is to this simple but innovative circuit. Power is supplied through switch S1 across monitor LED (D2) and limiter resistor, R2 (4.7 kohm). Capacitors C1 (0.1 µF) and C2 (100 µF) provide smoothing.

Construction Tips

Once you've chosen your assembly board, it's a good idea to use a 14-pin DIL socket for this circuit. As mentioned above, take special note of the power and ground pin locations—I would even go so far as to mark clearly on the board, the polarities for the supply feeds going into pin #4 (power) and pin #11 (ground). This is about the only critical part of the circuit layout.

The potentiometer, VR1, is obviously an off-board component, because of its physical size. Because of the need to vary the potentiometer shaft, a project case is highly recommended. With the nine-volt supply used, the input signal can have any excursion up to the supply voltage, from zero to nine volts. The LED (D1) monitor can be board mounted if the assembly board is set up on the top surface of the project case. Otherwise mount the LED on the front panel next to the input jack socket (J1).

Test Setup

Connect a suitable (slow-moving dc and amplitude-matched) signal source to the input and make sure the power is applied to the threshold detector before the signal source is switched on. This is a general rule when coupling together several circuits that have separate power supplies. Start with the last (this being the one at which the final output exists), and work backward toward the source. Generally you'll have just two such circuits coupled. Predetermine what the signal amplitude will be like, either by using a known source or an oscilloscope to verify the waveform. If you do have the luxury of an oscilloscope, it will be handy to monitor the input and watch for the point at which the output signals switches state. Start with VR1 set to a mid-point and work from there.

Parts List

Semiconductor
    IC1: LM 324 quad operational amplifier
Resistors (all resistors are 5 percent 1/4 W)
    R1: 1 kohm
    R2: 4.7 kohm

Capacitors
(All nonpolarized capacitors disc ceramic)
(All electrolytic capacitors have a 25 V rating)
    C1: 0.1 µF
    C2: 100 µF
Additional materials
    VR1: 100 kohm potentiometer
    D1: LED
    D2: LED
    J1: miniature 1/8-inch jack socket
    S1: single-pole, single-throw miniature switch
    Power supply: nine-volt battery

Project 6: Hi-Lo Trip Detector

Introduction

Having cut our teeth on the use of the single LM 324 operational amplifier, let's now look at using two of these in the hi-lo trip detector. It's easy enough to say you can just add another operational amplifier, and this is a technique commonly used in many books, but I always like to see the full details given for a circuit. In that way if the circuit doesn't work, you have a circuit to work with rather than wondering where you've gone wrong. In this circuit, we have this time two independent threshold-level detectors, which we'll call the upper and lower or hi-lo trip limits, to produce a dc voltage-range indicator. As the input voltage makes its positive-going excursion, one LED will show you when the first threshold level is crossed, and the second LED will alert you when the second next-higher threshold is crossed. Two potentiometers are this time wired up in a potential-divider mode (using the three terminals) in order to set the lower and higher trip levels.

Circuit Description

The first difference you'll notice with this circuit is that only one of the 1/4 operational amplifiers has a power and ground connection. That's because the operational amplifiers are in the same physical IC package. If we were using discrete LM 741s instead, there would be, of course, a power and ground connection to each operational amplifier. Both amplifiers are configured identically, with the input signal fed to both of their inputs as seen in Figure 5-5.

The common input signal drives the noninverting inputs, and separate LED monitors are set up across the two outputs to show when the thresholds are exceeded. The first operational amplifier, IC1a (1/4 LM 324) will be our lower-threshold section. Potentiometer VR1 (100 kohm) draws a minimum amount of current from the voltage source, which is a standard nine-volt battery. The wiper terminal of VR1 produces the variable reference voltage,

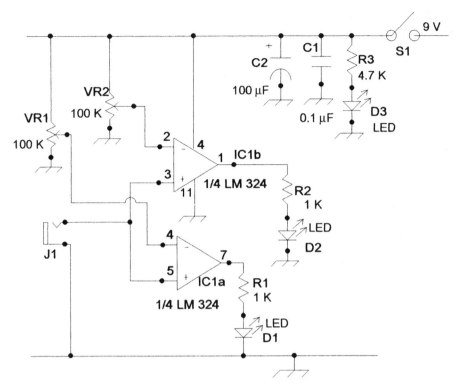

Figure 5-5 Hi-Lo Trip Detector

ranging from zero to the full supply voltage. The voltage across the wiper and ground terminals can be measured with a dc voltmeter. As an arbitrary value, set the low reference value to a convenient value, like 1 volt. The second half of the circuit has IC1b (1/4 LM 324) as the comparator, with VR2 (100 kohm) this time used for the high-voltage reference setting. A convenient value could be two volts. Needless to say, the circuits could be interchanged for the high-low thresholds, as they're identical. The usual LED monitors (D1 and D2) across the outputs (pin #7, pin #1) will light up when the signal exceeds the set limit thresholds. There are the standard current-limiting resistors, R1 (1 kohm) and R2 (1 kohm), associated with D1 and D2. Power is from a nine-volt battery fed in through supply switch S1. LED (D3) and resistor R3 (4.7 kohm) provide indications when the power is supplied to the circuit. Capacitors C1 (0.1 µF) and C2 (100 µF) smooth out the supply voltage.

Construction Tips

This circuit is built along the lines of the previous project and should be easy to follow, as the two circuit halves are identical in form. The components will be somewhat more packed on the assembly board, as there are twice as many components as the previous circuit.

Test Setup

Follow the same test routine as before. Your signal source will the same, except that you need to have a signal excursion in excess of the higher trip voltage, to get the LEDs to light up. It's not critical. An oscilloscope, again, is very useful as a monitor of the original signal's amplitude.

Here's a novel thought about this circuit, though: once you have verified the correct operation of this circuit, the potentiometers can be calibrated. Do this by putting a mark against the potentiometer controls, perhaps in 0.5-volt increments. When you've done this, this circuit will eliminate the need for an oscilloscope, as you can now use the LEDs to determine roughly the amplitude of any unknown signal. Let's say you have an unknown signal being applied. By adjusting the potentiometer (VR2) you can determine the maximum signal level (it works best if the signal is smooth and repetitive, like a sine wave). Should the signal be offset from the zero voltage level, the potentiometer, VR1, can be used to determine the lower voltage point.

Parts List

Semiconductors
    IC1a: 1/4 LM 324 quad operational amplifier
    IC1b: 1/4 LM 324 quad operational amplifier
Resistors (all resistors are 5 percent 1/4 W)
    R1: 1 kohm
    R2: 1 kohm
    R3: 4.7 kohm
Capacitors
(All nonpolarized capacitors disc ceramic)
(All electrolytic capacitors have a 25 V rating)
    C1: 0.1 µF
    C2: 100 µF
Additional materials
    VR1: 100 kohm potentiometer
    VR2: 100 kohm potentiometer
    D1: LED
    D2: LED
    D3: LED
    S1: single-pole, single-throw miniature switch
    Power supply: nine-volt battery

Project 7: LED Pulse Stretcher

Introduction

We're used to seeing an integrated circuit used as the heart of most circuits, but that need not always be the case. A few passive components and a lot of ingenuity can provide you with a nice, novel little circuit. LEDs are often used as a simple means of detecting slow-moving positive pulses. In order for the LED to light up, though, the pulse input must be of sufficient width to allow it to come on. Narrow or low-duty-cycle pulse widths will not light up the LED. This circuit produces a little stretching of the pulse, so the net effect is that the LED is given more time to light up, meaning, of course, that you can now see it. That's quite useful—and you don't even need an IC to do this! Of course, there are some restrictions, one being that the pulse has to have a low repetition rate.

Circuit Description

The heart of this circuit, shown in Figure 5-6, is based on the charging properties of a capacitor. When a dc voltage is applied to a capacitor, the capacitor builds up a charge until the full applied voltage exists across the capacitor even when the voltage is removed. The higher the capacitor value, the longer will be the voltage-retention capability.

We're using here a large capacitor C1 (47 µF) as our vehicle for storing our charge. Any charging circuit is composed of a capacitor and a resistor. Their product is called the "RC time constant." The higher this value, the more time the capacitor takes to charge up to the applied dc voltage. In the circuit shown, capacitor C1 has a value of 47 µF and resistor R1 a value of 47 ohm. The RC product (in units of seconds) is thus: $47.10^{-6} \times 47 = 0.002$ seconds, or 2 milliseconds. Use of the low value for resistor R1 produces a short time constant, which is what we want. This means that before the next pulse arrives, the capacitor will have charged up. Diode D1 (1N4001) is a rectifier diode and is strategically placed in the circuit to prevent the charge on capacitor C1 from leaking

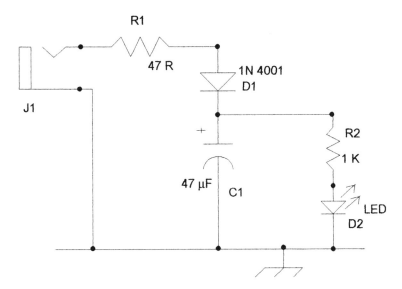

Figure 5-6 LED Pulse Stretcher

back into a low-impedance source. The rectifier conducts when a positive voltage is applied to the anode end. As the voltage builds up on the capacitor, the rectifier is no longer forward biased, and so the charge remains on the capacitor. The time constant for when the LED (D2) is on will be this time: $47 \times 10^{-6} \times 1000 = 0.047$ seconds, or 47 milliseconds. It seems a relatively larger number, but as far as the eye is concerned, it's a fairly short period. Resistor R2 (1 kohm) is the current limiter for D2.

Construction Tips

For once there's no need for the usual IC socket! There are five components to this build, and they're all passives! As you will need to couple in a signal, which'll probably come from a signal generator, I suggest a 1/8-inch miniature jack socket for the input signal. The pulse rate is so slow that there's no need for any special high-frequency considerations. Use a project case if you wish, but it hardly seems necessary here. The jack socket could be ingeniously anchored to the assembly board, which in itself need only be an inch or so on a side. As there's nothing to adjust, another alternative is to eliminate the board and just attach the components to each other—with point-to-point soldering. A small case, such as a film canister, makes an ideal housing. The LED indicator could be conveniently mounted at one end and the jack socket inserted into the other end.

Test Setup

Personally, I always like to verify a circuit's operation before committing to the final housing. You can easily lay everything out on the bench top and verify the circuit's function. A pulse generator having a variable frequency and adjustable duty cycle, as provided by a function generator, is ideal. Use an oscilloscope to determine what sort of waveform you're getting. Comparing the input pulse against the LED flashes will show you how the LED pulse stretcher is functioning. You'll see on the oscilloscope the narrow pulse dropping to zero, while the LED still stays on, because of the charging capacitor. If you've chosen a sufficiently low frequency for the signal input, the LED should die down before the next pulse comes along.

Parts List

Resistors (all resistors are 5 percent 1/4 W)
   R1: 47 ohm
   R2: 1 kohm
Capacitors
(All nonpolarized capacitors disc ceramic)
(All electrolytic capacitors have a 25 V rating)
   C1: 47 µF

Additional materials
D1: 1N4001 rectifier
D2: LED

Project 8: dc Level Shifter

Introduction

To close this section, here's another unusual circuit, one that you won't normally see featured—there's a coupling capacitor at the input but none at the output! It's a combination ac-plus-dc operational amplifier circuit that I've called a "dc level shifter." In regular ac amplifier circuits using the standard LM 741, there's the usual ac coupling capacitor at the input and another coupling capacitor at the output side, to remove the dc component. Using a single nine-volt supply voltage and starting off with a standard Vcc/2 biasing arrangement, we add a simple modification that will allow the ac output signal to be shifted above and below the Vcc/2 line, that is, above and below the nominal 4.5 volt value. There are some applications where you'll need a small adjustment to the superimposed dc bias voltage. If the ac signal amplitude is kept reasonably low in order to prevent overloading or clipping of the signal peaks, the circuit arrangement shown here will be very useful.

Circuit Description

Integrated circuit IC1 (LM 741) is the active circuit here for the schematic in Figure 5-7. The main amplifier section is configured as a regular inverting amplifier. The feedback resistor R2 (100 kohm) straddles the output terminal (pin #6) and the input terminal (pin #2). Together with the input resistor R1 (10 kohm), there's a gain of ×10 provided from the ratio of the resistors R2/R1. Capacitor C1 (0.1 µF) removes the dc component from the incoming signal. Note the absence of a capacitor at the output, since the intent is to have a dc signal available also at the output. Normally the biasing arrangement to the noninverting terminal (pin #3) requires a split Vcc/2 voltage. There's a similar-looking circuit here, except there's a three-resistor series arrangement. We start with resistor R3 (47 kohm) that's connected to the positive Vcc supply. R3 in turn is connected to a potentiometer, VR1 (50 kohm), wired as a two-terminal series resistor. VR1 in turn goes to R4 (27 kohm), which in turn gets terminated to the ground connection. The junction of R3 and VR1 goes to the noninverting terminal (pin #3) of IC1. When VR1 is adjusted to a value such that R3 equals VR1 + R4, the output signal sits exactly on the Vcc/2 line. With the aid of an oscilloscope you can see this quite easily. By varying the value of VR1, you can shift the ac signal above and below the reference Vcc/2 level.

Expensive commercial function generators have a variable dc offset capability, which you can inexpensively duplicate with this nice easy circuit.

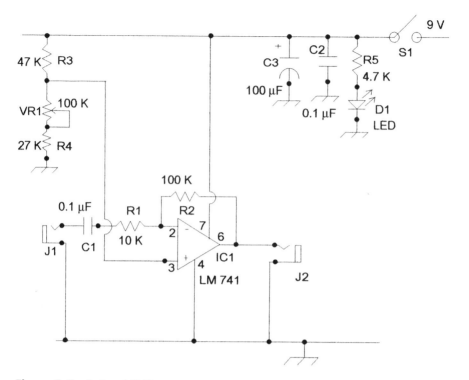

Figure 5-7 dc Level Shifter

This circuit's another example of making a simple change to a regular circuit (the inverting operational amplifier) to wind up with a really clever new circuit! That's the fun part of playing around with circuits and seeing what novel derivatives you can create.

Power is supplied to pin #7, and the ground connection is routed to pin #4. The output signal emerges from pin #6—without the usual coupling capacitor, as we want an ac + dc signal to appear there. The input can be coupled through a 1/8-inch miniature jack socket (J1), and make the connection from the output through another 1/8-inch miniature jack socket (J2). Switch S1 applies power to the circuit, which is monitored by LED D1 and resistor R5 (4.7 kohm). Capacitors C2 (0.1 µF) and C3 (100 µF) provide smoothing to the supply line.

Construction Tips

Start with positioning an eight-lead DIL socket for IC1 (LM 741) on your circuit assembly board of choice. As this circuit uses just one IC, I usually position the IC socket somewhere conveniently in the center of the board. Because of the inclusion of a number of mechanical components in the circuit, the potentiometer (VR1), input jack socket (J1), and output jack socket (J2) should be mounted on a project case for stability. The power switch (S1) and power-on LED indicator (D1) are also panel-mounted components. For an extra touch, potentiometer VR1 can be calibrated. Use a pointer knob for VR1 and mark off the center position where you have exactly a Vcc/2 point. This can be determined by measuring the dc voltage at pin #3 for different setting of VR1. When VR1 reaches one end of its travel, you've essentially got a potential-divider circuit, with the "top" resistor equal to 47 kohm (this is our value for R3) and the "lower" resistor equal to 77 kohm (this is our value for VR1 of 50 kohm plus the value of 27 kohm for R4). In that position you've got about 60 percent of the supply voltage appearing. When the opposite end of VR1's travel is reached, we now have the "top" resistor equal to 47 kohm (as before) and the "lower" resistor now equals 27 kohm (the value for R4), since VR1 equals zero ohms. Here we have about 36 percent of the supply voltage appearing across pin #3. So you see, we've got a nice swing either side of the 50 percent × Vcc value.

You can mark up the indicator knob in voltage steps (with 4.5 volts in the center position), but bear in mind that as the battery voltage drops, the values will no longer be an accurate representation. It's not really a problem, as it'll still give you an indication of which side of the Vcc/2 line you're biasing the output signal. Alternatively, the reference marks can be in terms of percentages, with 50 percent in the center. That way it will always be accurate, but to get the actual dc bias value, you'll have to do the calculation.

This is how you get the divided-down value. A basic potentiometer consists of two resistors in series; the upper one we'll call Rx, and the lower one we'll call Ry. The input is fed to the free end of Rx, and the free end of Ry is grounded. The junction between Rx and Ry is fed out. The percentage reduction of the input voltage that appears across the output is given by the ratio of Ry/(Rx + Ry).

Test Setup

This is one of the few circuits that requires an oscilloscope to be able to view the output waveform, as there are both ac and dc components in the signal output. An ac signal (sine wave or triangular wave) is required for the input feed. Ideally a function generator would be the best signal source, where you've got control over the signal amplitude. Any frequency can be used; a 1 kHz audio signal is ideal, and a peak-to-peak amplitude of a few tens on millivolts is fine. The amplifier has a gain of ×10, so you'd expect to see a 100 millivolt signal if a 10 millivolt signal feed is used. Start with the dc bias to the noninverting terminal (pin #3) set to Vcc/2. The signal output from pin #6 is fed to one of the oscilloscope's input channel. This input channel needs to be set to the dc range. Apply power to the dc level shifter circuit first, before switching on the function generator. What you should see on the monitored output channel is a clean, amplified version of the input signal. All being well, you can vary the potentiometer VR1 about the Vcc/2 value, and this should result in the output waveform being shifted about that reference value.

Parts List

Semiconductor
    IC1: LM 741 operational amplifier
Resistors (all resistors are 5 percent 1/4 W)
    R1: 10 kohm
    R2: 100 kohm
    R3: 47 kohm
    R4: 27 kohm
    R5: 4.7 kohm
Capacitors
(All nonpolarized capacitors disc ceramic)
(All electrolytic capacitors have a 25 V rating)
    C1: 0.1 µF
    C2: 0.1 µF
    C3: 100 µF
Additional materials
    VR1: 100 kohm potentiometer
    D1: LED
    J1: 1/8-inch miniature jack socket
    J2: 1/8-inch miniature jack socket
    Plastic project case
    Power supply: nine-volt battery

CHAPTER **6**

Schmitt Trigger Circuits: Projects 9–13

Introduction

The Schmitt trigger inverter (74LS14) is a versatile, digital, integrated circuit that transforms a slow-moving input signal to a fast-changing, glitch-free output signal and is capable of producing a variety of useful digital circuits. This section will feature a number of such circuit innovations, all very simple to build, and all unique to this IC. The key distinction of the Schmitt trigger is that it provides a clean, sharp-edged, square-wave pulse train. This TTL device comes in a 14-pin DIL package and runs off the usual five-volt supply requirements. The first circuit is a low parts-count pulse-train generator. The output waveform is automatically TTL compatible, so interfacing into a following TTL logic circuit works perfectly. If it were as widely available as the LM 555, it would be more commonly used as a universal square-wave generator, but the LM 555 is untouchable in that respect, especially with regard to the number of circuits incorporating the "555" timer IC. The 74LS14 Schmitt trigger has the inherent property, unlike the LM 555, of producing an almost perfect 50 percent duty cycle waveform that is retained even when the frequency is varied, unlike the LM 555's irritating property of a duty cycle that changes with frequency.

Project 9: Schmitt Pulse Oscillator

Introduction

A square-wave pulse-train generator forms the basis of a very useful circuit to have in the digital tool kit of circuit functions. By using the 74LS14 we can easily generate such a pulse train. Two timing components, a resistor and capacitor, are all that is needed to implement the pulse-train generation. The output pulse-train frequency is given approximately by $f = 1/(RC)$, where f is in units of hertz, R is in ohms, and C in farads. For example, the values $R = 10\,k\text{ohm}$ and $C = 0.1\,\mu F$ generate a convenient audio frequency of 1 kHz. The output waveform is a very good approximation to a 50 percent duty-cycle square wave.

Circuit Description

IC1 (74LS14) is a 14-pin device containing six (hex) separate Schmitt invertor circuits. One of these invertors is configured to function as an oscillator, as seen in Figure 6-1.

When the Schmitt trigger invertor is configured as a straight invertor, a positive-going signal input will translate to become a negative-going output signal, and vice versa. If you were to display the input and output waveforms on an oscilloscope, the phase difference would be clearly seen; when the input sits at a logic high, the output is at a logic low. With such a simple circuit (the invertor), it might at first glance be somewhat doubtful that there is anything more to be gained from this circuit. But there is more to this device than meets the eye at first. Apart from the IC itself, only two other components are needed to turn this device into an oscillator, and that's using just one of the six available invertors! Even the LM 555 requires more than two components to perform a similar task.

Resistor R1 (10 kohm) is coupled across the output (pin #2) and the input (pin #1). From pin #1, a capacitor C1 (0.1 µF) is connected to ground. These

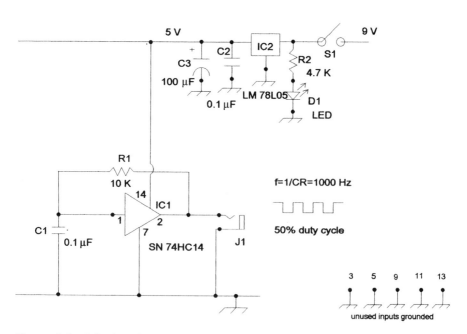

Figure 6-1 Schmitt Pulse Oscillator

two components (R1 and C1) are all that are needed for the simple timing calculation. The output signal is taken directly from pin #2 without a coupling capacitor (as the dc component has to be maintained). This is a regular TTL-compatible signal (zero volts at logic low and Vcc at logic high). The rest of the unused invertors need to be appropriately connected. All the unused outputs are left open circuit, but the unused inputs have to be grounded. The pins to be grounded are #3, #5, #9, #11, and #13. Grounding unused inputs of unused multidevices is a common practice with digital circuits, in order to provide circuit operating stability.

The signal output is coupled from a 1/8-inch miniature jack socket (J1). With a power requirement of five volts and a nine-volt battery, we have to step down the supply voltage. As seen in earlier chapters, the quickest, easiest, and most reliable way to do this is by using a five-volt regulator IC. Next to the ever-popular LM 555 and LM 741 ICs, the three-terminal regulator has got to be one of the most necessary in the digital tool kit. To complete the power-supply section, we have the usual power-on/off switch, S1, located on the feed side. IC2 is the five-volt (LM78L05) regulator. Feed in nine volts, and out comes a convenient five volts. LED (D1) and resistor R2 (4.7 kohm) form the power-on indicator circuit. Two supply capacitors C2 (0.1 µF) and C3 (100 µF) aid in the stability of the supply voltage.

Construction Tips

Start with a 14-pin DIL IC socket as your prime component. As there are so few components needed, the size of your assembly board can be dictated by the size of the socket itself. As is common with a lot of digital devices, the power pin (#14) is located at the upper left-hand corner and the ground pin (#7) at the lower right-hand corner. Bring out a direct connection to the upper positive five-volt line from pin #14. Along the lower end of the IC, bring out a connection from the ground, pin #7. Then bring out the rest of the unused inputs to ground. Pins #3 and #5 are located along the lower end of the IC, which is convenient. Pins #9, #11, and #13 are located along the top of the IC. Depending on your board arrangement, you could have a common ground rail along the upper edge of the IC. Use that if you do. If not, bring out links to the lower ground rail. As with any circuit that incorporates a mechanical device, such as a switch, potentiometer, or socket, it's highly advisable to use a project case (plastic's great) to house the build. The battery is probably going to be the largest component, so choose the case with that in mind.

Test Setup

Testing is nothing more than switching on. However, access to an oscilloscope is highly recommended here, as it's the only way to observe the approximately 50 percent duty-cycle property of the output square-wave form. The edges of the waveform should be razor sharp—a characteristic of the Schmitt.

Parts List

Semiconductors
    IC1: 74LS14 TTL Schmitt trigger
    IC2: LM78L05 regulator
Resistors (All resistors are 5 percent, 1/4 W)
    R1: 10 kohm
    R2: 4.7 kohm
Capacitors
(All nonpolarized capacitors disc ceramic)
(All electrolytic capacitors have a 25 V rating)
    C1: 0.1 µF
    C2: 0.1 µF
    C3: 100 µF
Additional materials
    D1: LED
    J1: 1/8-inch miniature jack socket
    S1: single-pole, single-throw miniature switch
    Power supply: nine-volt battery

Project 10: Turn-On Delay Schmitt

Introduction

The Schmitt trigger's circuit versatility shows to good effect in this unusual turn-on delay circuit. There could be an application where a circuit is required to turn on only after a certain delay period has elapsed. The core circuit shown can be easily integrated into a more extensive circuit chain.

Circuit Description

The Schmitt, being an invertor, has an antiphase relation between the output and input. This means that the output goes high when the input is low, and vice versa. Consider the state of the circuit shown in Figure 6-2, at the power switch on point in time when switch S1 is engaged to supply power to the circuit.

The critical input controller switch, S2, is in the closed position at the start of the proceedings. There are two resistors feeding from the Schmitt input terminal (pin #1): R1 (100 kohm) coupling the input to the Vcc rail and R2 (1 Mohm) going to ground. Resistors R1 and R2 form a basic potential divider circuit such that the potential existing at pin #1 is given by: Vin × [R2/(R1 + R2)] = 9 × ($1.10^6/1.1.10^6$) = 8.2 volts. As this is an invertor circuit, the expected logic output state is a logic low when the input is fed with a high-level input signal. As long as this equilibrium situation exists, there is no change in the output conditions; the circuit output is essentially in the off state.

Schmitt Trigger Circuits: Projects 9–13   **101**

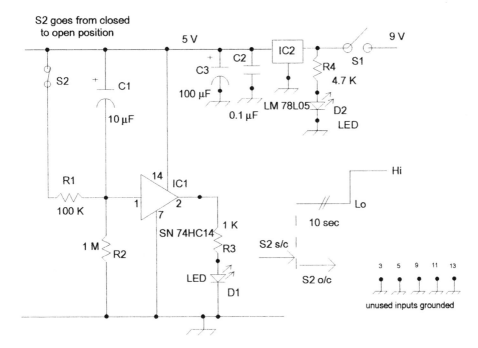

Figure 6-2   Turn-On Delay Schmitt

To effect the turn-on delay, switch S2 is changed to the off position. Resistor R1 is now out of the circuit. Capacitor C1 (10 µF), coupling pin #1 to the positive supply voltage rail, now comes into play. A charging time constant now takes place, involving C1 and R2. The charging time constant is given by the product $C1 \times R2 = 10.10^{-6} \times 1.10^{6} = 10$ seconds. Time constants always have units of time (in seconds); R (resistance) is in units of ohms, and C (capacitance) is in farads. Ten seconds after the point S2 is switched from a closed to open position, the output will turn on. A LED (D1) and current-limiter resistor, R3 (1 kohm), provides an easy monitor for the output waveform.

The five-volt supply requirements are taken care of with the regulator circuit, IC2 (LM78L05), as shown in the positive power-supply rail. The usual power-on switch S1 is positioned at the input side of the regulator IC2, with LED (D2) acting as the power-on indicator. Resistor R4 (4.7 kohm) provides current limiting for D2. Two smoothing capacitors, C2 (0.1 µF) and C3 (100 µF), are located at the downside of IC2.

Construction Tips

IC1 is located on the assembly board by means of a 14-lead DIL IC socket. Because of the inclusion of mechanical components, e.g., the switches (S1 and S2) and LEDs (D1 and D2), a project case is recommended for the build. Other than the IC itself, there are so few additional components that the build should be relatively straightforward. IC2 is a three-terminal device that is directly board-mounted. You can easily verify the operation of this device by placing the input terminals of IC2 across a nine-volt battery and measuring across the output and ground terminals; confirm that you do in fact get a five-volt positive output. Note that IC2 is a positive-input device only and will function correctly only when the input feed voltage has a positive polarity. It's a good idea to mark the switch position for S2, with a legend for the short circuit and the open circuit positions, as in that way there is no confusion as to how to engage the turn-on delay function.

Test Setup

The test setup is simple. No other additional pieces of test equipment are needed, as the LED (D1) will monitor the delay between when the time delay-switch S2 is switched from the on to off position, and the time D1 turns on. Remember that delay-switch S2 must be kept in the on position when it's switched from the initial on-to-off state. This is so that the capacitor C1 can be charged up.

Parts List

Semiconductors
    IC1: 74LS14 TTL Schmitt trigger
    IC2: LM78L05 regulator
Resistors (All resistors are 5 percent, 1/4 W)
    R1: 100 kohm
    R2: 1 Mohm
    R3: 1 kohm
    R4: 4.7 kohm
Capacitors
(All nonpolarized capacitors disc ceramic)
(All electrolytic capacitors have a 25 V rating)
    C1: 10 µF
    C2: 0.1 µF
    C3: 100 µF
Additional materials
    D1: LED
    D2: LED
    S1: single-pole, single-throw miniature switch
    S2: single-pole, single-throw miniature switch
    Plastic project case
    Power supply: nine-volt battery

Project 11: Turn-Off Delay Schmitt

Introduction

This Schmitt trigger circuit variation is a direct complement to the previous circuit. The turn-off delay is this time affected by a modified front end of the earlier circuit. Otherwise the rest of the circuit stays the same.

Circuit Description

IC1 is the 74LS14 Schmitt trigger, as seen in Figure 6-3. The critical timing components are switch S2, resistor R1 (100 kohm), resistor R2 (1 Mohm), and capacitor C1 (100 µF). Switch S2 is a regular single-pole, single-throw toggle switch that normally sits in the closed position. In that position (with S2 short circuited), the input to the Schmitt trigger is a potential divider composed of R2 (1 Mohm) in the upper arm and R1 (100 kohm) in the lower arm. From a calculation of the voltage appearing, we get: $V_{in} \times [R1/(R1 + R2)] = 9 \times (100.10^3/1.1.10^6) = 0.82$ volts. The Schmitt being an invertor circuit, we would expect to see the output go high under these conditions, and of course we do, as indicated by the LED (D1) turning on. Resistor R3 (1 kohm) is the current-limiting resistor for D1.

Switch S2 initially is in the closed position, to set up the equilibrium conditions. At the point of S2 being turned to the off position, the following event

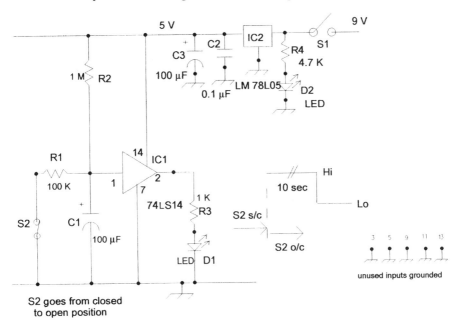

Figure 6-3 Turn-Off Delay Schmitt

takes place: resistor R1 is switched out of the circuit. Capacitor C1 (100μF) takes over, together with resistor R2 (1 Mohm), to charge according to the time-constant product of C1 × R2 = $10.10^{-6} \times 1.10^{6}$ = 10 seconds.

The time constant is in units of time (in seconds), R (resistance) is in units of ohms, and C (capacitance) is in units of farads. Ten seconds after the point S2 is switched from a closed to open position, the output will turn off. A LED (D1) and current-limiter resistor, R3 (1 kohm), provides a monitor for the output waveform.

The five-volt supply requirements are taken care of with the regulator circuit, IC2 (LM78L05), as shown in the positive power-supply rail. The power-on switch S1 is positioned at the input side of the regulator IC2, with LED (D2) acting as the power-on indicator. Resistor R4 (4.7 kohm) provides current limiting for D1. Two smoothing capacitors, C2 (0.1 μF) and C3 (100 μF), are located at the downside of IC2.

Construction Tips

IC1 is located on the assembly board by means of a 14-lead DIL IC socket. Because of the inclusion of mechanical components, for example, the switches (S1 and S2) and LEDs (D1 and D2), a project case is recommended for the build. Other than the IC itself, there are so few additional components that the build should be relatively straightforward. IC2 is a three-terminal device that is directly board mounted. You can easily verify the operation of this device by placing the input terminals of IC2 across a nine-volt battery, and measuring across the output and ground terminals; confirm that you do in fact get a five-volt positive output. Note that IC2 is a positive-input device only and will function correctly only when the input feed voltage has a positive polarity. It's a good idea to mark the switch position for S2, with a legend for the short circuit and the open circuit position, as that way there is no confusion as to how to engage the turn-on delay function.

Test Setup

The test setup is simple. No other additional pieces of test equipment are needed, as the LED (D1) will monitor the delay between when the time delay-switch S2 is switched from the on to off position, and the time D1 turns on. Remember that delay-switch S2 must be kept in the on position when it's switched from the initial on-to-off state. This is so that the capacitor C1 can be charged up.

Parts List

Semiconductors
IC1: 74LS14 TTL Schmitt trigger
IC2: LM78L05 regulator

Resistors (All resistors are 5 percent, 1/4 W)
  R1: 100 kohm
  R2: 1 Mohm
  R3: 1 kohm
  R4: 4.7 kohm
Capacitors
(All nonpolarized capacitors disc ceramic)
(All electrolytic capacitors have a 25 V rating)
  C1: 10 µF
  C2: 0.1 µF
  C3: 100 µF
Additional materials
  D1: LED
  D2: LED
  S1: single-pole, single-throw miniature switch
  S2: single-pole, single-throw miniature switch
  Plastic project case
  Power supply: nine-volt battery

Project 12: Schmitt Triangle Generator

Introduction

There are many audio test applications that would benefit from the services of a smoothly transitioning oscillator, such as a sine wave. Sine wave oscillators are unfortunately not as simple to put together as square-wave generators. The square wave is too harsh a waveform to be able to discern any distortions generated by an audio amplifier, rather than from the test input signal. However, it is a simple matter to convert a square wave into a triangle wave, and a triangle-wave signal makes a pretty good approximation to a sine wave, especially as far as the audio tone quality is concerned. Triangle waves are a useful signal source, as they have a changing voltage characteristic that is more easier viewed than the sharply transitioning square wave. A triangle waveform rises linearly from a minimum to a maximum peak and repeats thereafter. As many hobbyists include both digital and analog circuits in their project builds, this circuit is a nice interface between the two. As this is a chapter devoted to the Schmitt trigger, the circuit shown here is based on that device. An oscilloscope will allow you to view the triangle waveform generated.

Circuit Description

IC1 (74LS14) is the Schmitt trigger, where one of the six available invertor circuits are used. The circuit is shown in Figure 6-4. The frequency output is a fairly symmetrical square wave, for later conversion into a good approximation to a triangle wave. A feedback resistor R1 (10 kohm) is wired across the

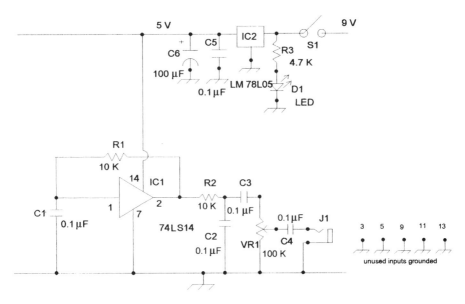

Figure 6-4 Schmitt Triangle Generator

output terminal (pin #2) back to the input terminal (pin #1). From input terminal (pin #1), a capacitor C1 (0.1 µF) goes to ground. The output pulse-train frequency is approximately given by: $f = 1/(RC)$, where f is in units of hertz, R is in ohms, and C in farads. Using the values of R = 10 kohm and C = 0.1 µF generates a convenient audio frequency of 1 kHz, from: $f = 1/(10.10^3 \times 0.10^{-6}) = 1/0.001 = 1000$ Hz. We can conveniently increase the frequency by decreasing either the timing components, resistor R1, or the capacitor C1. The output waveform is a very good approximation to a 50 percent duty-cycle square wave. The balance of the five invertors' unused input terminals have to be grounded in order to get the circuit to operate properly, so ground pins #3, #5, #9, #11, and #13. The output pins just remain open circuit.

From the output terminal, two components, resistor R2 (10 kohm) and capacitor C2 (0.1 µF), form a simple integrator circuit. When an integrator is fed with a square wave, the output becomes a triangle wave. The integrator RC combination limits the high-frequency response of the output waveform and at the same time applies some attenuation to the final signal.

At a frequency of 1 kHz, the integrator capacitor C2 (0.1 µF) has an impedance of 1.59 kohm. Integrator resistor R2 is a 1 kohm resistor, so as a simple potential-divider network, we'd see about a 50 percent reduction in the signal amplitude. As the original signal has an amplitude close to the supply voltage of five volts, the integrator output would be around half that (approximately 2.5 volts).

For added versatility in audio test applications, the final three components provide for a continuous reduction of the output waveform. Audio applications require a dc-free component; capacitor C3 (0.1 µF) serves this purpose. Potentiometer VR1 (100 kohm) provides the amplitude control, and, finally, capacitor C4 (0.1 µF) provides the ac coupling into the output jack socket (J1). Because the generated output is deliberately limited to one frequency, we don't have the problem of the output amplitude from the integrator varying with frequency. That means that the potentiometer can be conveniently graduated with a marker for the output amplitude. Furthermore, the circuit component values for the integrator are optimized for the single frequency of the incoming waveform. Different incoming frequencies will cause departures from the ideal converted waveform, hence this circuit is deliberately limited to a fixed-frequency configuration.

The signal output is coupled from a 1/8-inch miniature jack socket (J1). With a power requirement of five volts and a nine-volt battery, we have to step down the supply voltage. The quickest, easiest, and most reliable way to do this is by using a five-volt regulator IC2 (LM78LO5). To complete the power-supply section, we have the usual power-on/off switch, S1 (single-pole, single-throw), located on the feed side of IC2. Couple in nine volts, and out comes a convenient five volts. LED (D1) and resistor R3 (4.7 kohm) form the power-on indicator circuit. Two supply capacitors, C5 (0.1 µF) and C6 (100 µF), aid in the stability of the supply voltage.

Construction Tips

IC1 is located on the assembly board by means of a 14-lead DIL IC socket. Because of the inclusion of mechanical components, for example, the switch (S1), jack socket (J1), potentiometer (VR1), and LED (D1), a project case is recommended for the build. Other than the IC itself, there are so few additional components that the build should be relatively straightforward. IC2 is a three-terminal device that is directly board mounted. The input terminal is connected to the positive 9 V supply. The output terminal supplies the 5 V. IC2 is a positive-input device only and will function correctly only when the input feed voltage has a positive polarity.

Test Setup

Testing is nothing more than switching on. However, access to an oscilloscope is highly recommended here, as it's the only way to observe the triangle-wave property of the triangle wave output. Vary the potentiometer (VR1) control and confirm that the amplitude varies smoothly from zero to a maximum. It's a good idea to put down a graduated marker against the control potentiometer (VR1), perhaps in 100 millivolt peak-to-peak amplitude steps. That way, you'll have a more useful signal injector for audio amplifier testing.

Parts List

Semiconductors
    IC1: 74LS14 TTL Schmitt trigger
    IC2: LM78L05 regulator
Resistors (All resistors are 5 percent, 1/4 W)
    R1: 10 kohm
    R2: 1 kohm
    R3: 4.7 kohm
Capacitors
(All nonpolarized capacitors disc ceramic)
(All electrolytic capacitors have a 25 V rating)
    C1: 0.1 µF
    C2: 0.1 µF
    C3: 0.1 µF
    C4: 0.1 µF
    C5: 0.1 µF
    C6: 100 µF
Additional materials
    VR1: 100 kohm potentiometer
    D1: LED
    J1: 1/8-inch miniature jack socket
    S1: single-pole, single-throw miniature switch
    Power supply: nine-volt battery

Project 13: Schmitt Switch Debouncer

Introduction

Unlike some circuits where the integrated circuit itself is not particularly critical and could be replaced with another generic type (e.g., replacement of a LM 741 op amp with a LM 324 op amp), this novel circuit relies specifically on the characteristic unique to the Schmitt trigger itself. Ordinary mechanical switches are notorious for generating signal transients at the point of crossover (from switch-on to switch-off, and vice versa). For straight dc switching applications where there can be a long time-lag between switch transitions, mechanical switches are perfectly suited to their intended use. In logic circuits, inputs commonly are taken to high/low logic levels to trigger downstream logic changes. Mechanical switches at first glance seem the easiest way to implement these transitions, but the inherent transients generated make their direct use unsuitable; beginners to logic circuits will encounter no end of headaches in trying to troubleshoot apparently faulty circuits. Transients will cause false triggering, leading to unpredictable logic states. These switching actions have to be cleaned up electrically—the solution is to use a switch debouncer, and that's where the Schmitt comes in!

Circuit Description

The most called-for application for the Schmitt debouncer will be in a situation where the output goes high when a mechanical switch is closed, that is, the Schmitt moves in phase with the switch action. The circuit is shown in Figure 6-5.

IC1 is the 74LS14 Schmitt trigger, connected with a resistor R1 (100 kohm) going from input pin #1 and Vcc. From the same terminal a capacitor C1 (0.01 µF) is connected to ground. The time constant for this combination is a fairly quick 1 millisecond activation time, from the time-constant product of: $C1 \times R1 = 0.01 \cdot 10^{-6} \times 100 \cdot 10^{3} = 1$ millisecond. Switch S1 is connected between pin #1 and ground. If you've been following the earlier Schmitt trigger circuits in this chapter, take special note of the different component values used in the input section. It is important to stay with the values given with each circuit. The output will change instantly (for all intents and purposes) to a logic high as soon as the switch S1 is closed. When S1 is opened, the output drops instantly to zero. The important difference with this Schmitt circuit is that we're looking at the enhanced output with reference to the input signal. In all of the previous circuits, there was no concern regarding enhancement between the output and input. There is a monitor, LED (D1), and associated resistor, R2 (1 kohm), across the output as a useful indicator of the logic state.

The power requirement for this circuit is five volts, and using a nine-volt battery means we have to step down the supply voltage. This is done by using a five-volt regulator IC2 (LM78LO5). To complete the power-supply section, we have the usual power-on/off switch, S2 (single-pole, single-throw), located on

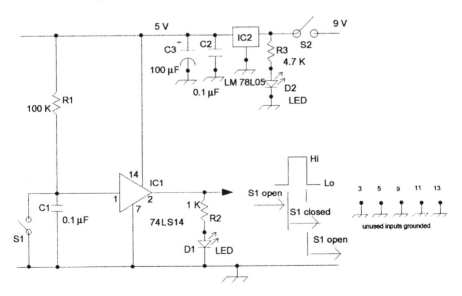

Figure 6-5 Schmitt Switch Debouncer

the feed side of IC2. LED (D2) and resistor R3 (4.7 kohm) form the power-on indicator circuit. Two supply capacitors, C2 (0.1 µF) and C3 (100 µF), aid in the stability of the supply voltage.

Construction Tips

This circuit has more applications with a follow-on circuit than as a stand-alone demonstration circuit, so a project case is more than just a requirement for containing the mechanical components. Switch S1 needs to be clearly marked on the front panel with the "open circuit" and "short circuit" positions. The output monitor will also clue you in as to the state of the input switch S1, but personally I prefer to have S1 marked up also. The balance of the five invertors' unused input terminals have to be grounded in order to get the circuit to operate properly, so ground pins #3, #5, #9, #11, and #13. The output pins just remain open circuit.

Test Setup

Testing's a little difficult, as the output pulse is going to look like any regular TTL pulse. The glitches generated as a result of S1's toggle positions are also not directly viewable, as they're much too fast. So the only test setup that can be effective is to use the output as a feed source for enabling the logic states for follow-on devices.

Parts List

Semiconductors
    IC1: 74LS14 TTL Schmitt trigger
    IC2: LM78L05 regulator
Resistors (All resistors are 5 percent, 1/4 W)
    R1: 100 kohm
    R2: 1 kohm
    R3: 4.7 kohm
Capacitors
(All nonpolarized capacitors disc ceramic)
(All electrolytic capacitors have a 25 V rating)
    C1: 0.1 µF
    C2: 0.1 µF
    C3: 100 µF
Additional materials
    D1: LED
    D2: LED
    J1: 1/8-inch miniature jack socket
    S1: single-pole, single-throw miniature switch
    S2: single-pole, single-throw miniature switch
    Power supply: nine-volt battery

CHAPTER **7**

Versatile Digital ICs: Projects 14–20

This chapter introduces you to a new variety of logic integrated circuits (ICs) that I've found particularly useful when starting to work with digital circuits. The circuits start simple and then increase a little in complexity. There is an almost endless list of digital devices available, and deciding where to begin is almost a task in itself (unlike the ease of choice with analog ICs, such as the ubiquitous LM 741, LM 555, and LM 386). The selection of introductory logic circuits I've presented here will help cut through the confusion (especially if you're a newcomer to digital logic circuits). I've used these devices many times over the years, and they've proved not only easy to use, reliable, and predictable, but also useful in demonstrating practical logic functions.

This chapter opens with the introduction of one of the simplest basic logic gates, the MC 14001 NOR gate IC. This is a CMOS-series logic device operating between a wide Vcc range of 3 to 18 volts (unlike the more common five-volt supply for TTL circuits). When driven from a five-volt supply, CMOS logic can be directly interfaceable to standard TTL logic—a nice advantage of using CMOS. The reason for using CMOS logic is that some logic devices are more commonly available in the CMOS technology format, and in addition, it introduces you to working with the second most common logic technology family. The MC 14001 is a quad device (it has four separate devices, with two input NOR gates) that is housed in one 14-lead DIL package. A basic introduction into how logic gates function can be gained by working with this starter device. The purpose of this first project with the NOR gate is to see how the output changes with various combinations of the input logic state.

Project 14: NOR Gate Demo

Introduction

The NOR gate, one of the basic logic devices, is ideal for experimenting and gaining familiarity with the way logic devices operate. The basic

NOR demo circuit shown here uses mechanical switches to set the input logic states—but with a difference! As we've seen in an earlier chapter, mechanical switches produce electrical glitches that can result in false logic states. However, for circuit-demonstration simplicity, there's a way around this, albeit somewhat cumbersome. The power switch in this circuit has to be turned off before the input switches are reset—but as this is a demo circuit, the consequence is minimal. Take note of this precaution.

Circuit Description

The NOR gate demo project uses a quarter of the four independent two-input NOR gates contained in IC1 (MC 14001). The circuit is shown in Figure 7-1.

The operation of the NOR gate is in accordance with the truth table shown below. The output from the NOR gate goes high only when both of the inputs are taken high. Power to the device is supplied from a nine-volt battery. The MC 14001 is a CMOS device that will operate off a very wide supply-voltage range, from 3 to 18 volts. When the MC 14001 is used to drive a following TTL stage, all that is needed for logic compatibility is a five-volt supply instead for the MC 14001. CMOS logic circuits has the advantage of running off a wide supply-voltage range, the only necessary precaution being that standard antistatic handling precautions needs to be followed to prevent destruction of the device.

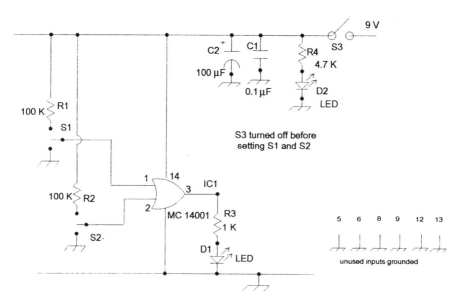

Figure 7-1 NOR Gate Demo

NOR Gate Truth Table

	Input #1	Input #2	Output
1)	0	0	1
2)	0	1	0
3)	1	0	0
4)	1	1	0

Since there are two inputs, four possible logic state combinations can exist: condition (1), input #1 low with input #2 low; condition (2), input #1 low with input #2 high; condition (3), input #1 high with input #2 low; and condition (4), input #1 high with input #2 high. To set up the input conditions, switches S1 and S2 (single-pole, single-throw types) are used to place the inputs (pins #1 and #2) into the high logic state, via resistors R1 (100 kohm) and R2 (100 kohm), which are taken to the Vcc rail. The reason is that the inputs (to logic circuits) cannot be left floating, as their logic states are then unpredictable, leading to an unreliable output states. The output (pin #3) logic state is monitored with a LED (D1). D1's current limit is provided through resistor R3 (1 kohm). Power is introduced to IC1's pin #14 from the output end of the power-supply switch S3. LED D2 and resistor R4 (4.7 kohm) monitor the presence of the supply voltage. Capacitors C1 (0.1 µF) and C2 (100 µF) provide smoothing of the supply line. The ground connection to IC1 is via pin #7. The output state will go high only when both of the inputs are taken to a logic low, and that's the aim of this circuit, to verify the truth table contents.

Construction Tips

A 14-lead DIL socket is recommended for this build. Take care when inserting the IC itself that you use a grounding strap to prevent any charge buildup from being discharged on the device itself. The MC 14001, being a CMOS device, is static sensitive. Also minimize any undue shuffling of your feet (one of the chief causes of static buildup) prior to picking up the MC 14001. A project case is essential, because of the three switches (S1, S2, S3) used in this demo circuit. Label each switch carefully—S1 (high/low position), S2 (high/low position), S3 (on/off position)—as their positions will be critical to verifying the NOR's truth table states. Switches S1 and S2 must be single-pole, double-throw types. Switch S1 is specified as a single-pole, single-throw type, but you can easily use the single-pole, double-throw type. So to cut down on your inventory of different types of switches, I'd suggest staying with the more universal single-pole, double-throw type. Use the soft-pressure types when buying these switches—that is, the type that requires the minimum of finger pressure to toggle—otherwise you're going to find that the project case housing will tip with the pressure needed to toggle an unnecessarily heavy-duty switch.

Test Setup

This test setup's different from the usual "switch on and monitor the output" process. Because of the inclusion of mechanical switches, it is essential that the following steps be adhered to. Mechanical switches are deliberately incorporated here, in spite of the previous precautions about switch-generated glitches. However, here we're side-stepping those issue, by making sure that the power switch S3 is always turned off before the input selector switches (S1 and S2) are set to the new positions. To toggle through the four input combination states, therefore, you're going to repeat this operation four times. Note that glitches are not guaranteed to occur with the input switch transitions, so you could still get the correct output state transitions, but you can be sure that under these conditions, the output will not be predictable—and as far as logic circuits are concerned, that cannot be tolerated.

Parts List

Semiconductor
  IC1: MC 14001 CMOS quad NOR gate
Resistors (all resistors are 5 percent, 1/4 W)
  R1: 100 kohm
  R2: 100 kohm
  R3: 1 kohm
  R4: 4.7 kohm
Capacitors
(All nonpolarized capacitors disc ceramic)
(All electrolytic capacitors have a 25 V rating)
  C1: 0.1 µF
  C2: 100 µF
Additional materials
  D1: LED
  D2: LED
  S1: single-pole, double-throw miniature switch
  S2: single-pole, double-throw miniature switch
  S3: single-pole, single-throw miniature switch
  Power supply: nine-volt battery

Project 15: SCR Latch Demo

Introduction

A solid-state latch has the unique property of remaining in a latched (conducting) state even after the trigger voltage has been removed. This circuit is an introduction to the next circuit, which is a cross-coupled NOR gate designed to simulate a latch. The circuit operation of the SCR (silicon-controlled rectifier) latch is simpler to follow than that circuit.

Circuit Description

The latch circuit centers around the intrinsic property of SCR1 (MCR 106-6). A latching circuit is defined as one that upon receipt of a trigger signal is converted from an inactive or nonconducting state into an active or conducting state and will remain in the converted active state even when the trigger source is removed. This is unlike the condition of many logic states, whose output condition is dependent on the input stimulus remaining in place; remove the input stimulus, and the output will change (as seen in the earlier NOR demo circuit). With that in mind let's take a look at the circuit shown in Figure 7-2.

The SCR is a three-terminal device, resembling a diode with an extra terminal. The anode and the cathode are the more customary terminals; they resemble diodes. A center terminal, called the gate electrode, is the additional trigger control. The SCR is connected into the circuit, with the anode coupled to the positive supply rail (similar to a NPN transistor circuit) and the cathode end grounded. The external load to be controlled by the SCR is connected in series with the anode. The SCR is normally in the nonconducting state, that is, no current flows through the device. When the SCR conducts, a high current flows through the device (the SCR), thus connecting the load across the supply rails. Resistor R1 (1 kohm) is inserted in series with the gate electrode and acts as a current limiter for the gate electrode. The value is not critical, and the resistor can be removed if extra sensitivity is required or where the trigger source is voltage or current limited. A potentiometer, VR1 (100 kohm), coupled across the supply rails, acts as a variable voltage trigger source for the gate electrode. Switch S1 is a series feed between VR1's wiper terminal and

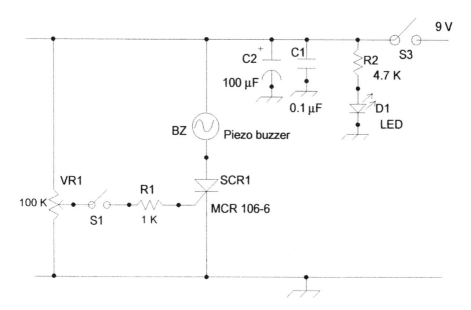

Figure 7-2 SCR Latch Demo

input resistor R1. A general-purpose piezo buzzer, BZ, can be used to act as a load in the SCR's anode.

The circuit operates in the following sequence. Switch S1 must be in the off position and VR1 set to the zero voltage value (this can be determined by connecting a dc voltmeter across VR1's wiper terminal and ground) before power is applied. Power switch S2 then applies power to the circuit. Switch S1 is closed, and the trigger voltage is incremented through VR1. Only a small positive voltage is required to trigger the SCR into a conducting state, indicated by the piezo buzzer turning on. Switch S1 can be opened and the SCR will still remain latched. To reset (turn off) the latch, turn power switch S2 off. Turning power switch S2 on again will leave the latch in the off state. To re-energize the latching action, switch S1 needs to be turned on again.

The gate terminal is very sensitive to even a small value of positive voltage; even positive pulse transients will trigger the latch (the gate trigger does not have to be a static dc voltage). Hence, this simple device is an ideal detector of fast, random, positive signal pulses. For example, if you're looking to verify the occurrence of a random noise pulse in a circuit, this SCR latch will allow you to do so without having to watch the circuit indefinitely. An LED acting as the monitor will turn on and remain on as evidence of the event having taken place.

Construction Tips

With so few components needed in this circuit, you're only going to need a very small assembly board to mount the components on. Because of the mechanical switching action, a project case is necessary to house the mechanical parts. It's also a good idea to calibrate the potentiometer with some reference marks for voltage being applied to the gate terminal. In that way you can see how small a voltage is needed to enable the circuit.

Test Setup

1. Toggle switch S1 to the off position and set potentiometer VR1 to the zero volts position
2. Toggle switch S2 to the on position
3. Toggle switch S1 to the on position and slowly advance VR1 till the device latches.
4. Toggle switch S1 to the off position; the latch is still enabled.
5. To reset (turn the latch off), toggle switch S2 to the off position.
6. To start the sequence again, toggle S2 and then S1 on.

Parts List

Semiconductor
    SCR1: MCR 106-6 silicon-controlled rectifier

Resistors (all resistors are 5 percent, 1/4 W)
    R1: 1 kohm
    R2: 4.7 kohm
Capacitors
(All nonpolarized capacitors disc ceramic)
(All electrolytic capacitors have a 25 V rating)
    C1: 0.1 µF
    C2: 100 µF
Additional materials
    BZ: general-purpose piezo buzzer
    VR1: 100 kohm potentiometer
    D1: LED
    S1: single-pole, single-throw miniature switch
    S2: single-pole, single-throw miniature switch
    Power supply: nine-volt battery

Project 16: NOR Gate Latch

Introduction

The previous circuit using the discrete SCR latch will have introduced you to the fundamental aspects of a latching circuit. This circuit shows how the NOR gate can be configured as a more versatile latch, illustrating that basic logic gates can provide more comprehensive functions than just the nominal gate function.

Circuit Description

This circuit uses two of the NOR gates contained within the MC 14001. Each NOR gate is a two-input device. IC1a and IC1b are each 1/4 of the MC 14001. IC1a is the upper NOR gate in the circuit. The output from pin #3 is cross-coupled back to the input (pin #5) of IC1b and in a similar manner the input from pin #2 of IC1a is cross-coupled to the output of IC1b (pin #4). The output from the latch (which is the combination of the two cross-coupled NOR gates) is taken from pin #4 of IC1b. The second input for IC1b is taken to ground via a resistor R3 (100 kohm). A capacitor C2 (0.047 µF) across resistor R3, inhibits any spurious responses in the circuit. A similar setup exists for IC1a. A resistor R2 (100 kohm) goes from input of IC1a (pin #1) to ground and capacitor C1 (0.047 µF) provides the same stability against spurious responses upsetting the circuit. The schematic is shown in Figure 7-3.

The next section of the circuit is a little different. A high value resistor R1 (1 Mohm) is taken from the input pin #1 of IC1a to Vcc. This high value for R1 ensures that the quiescent current in this circuit is very low, around 9 microamp. The quiescent current is the current drawn by the circuit in its off state. This circuit operates off a nine-volt battery, and as a latch it is going to be

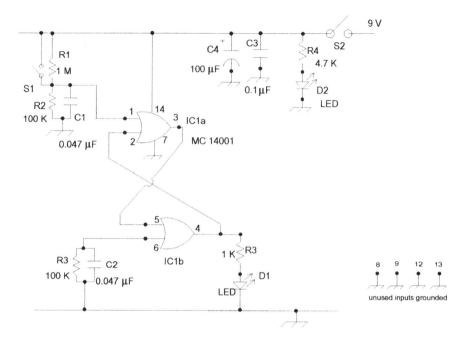

Figure 7-3 NOR Gate Latch

continually on for an indefinite period. Hence to minimize the current drain R1 should be a large value. Resistors R1 and R2 form a potential divider circuit to bias the input to IC1a at a logic low. With the values specified for R1 (1 Mohm) and R2 (100 kohm), 0.8 volts are applied. This is a logic low level. R1 could even be increased to 10 Mohm for even further saving in quiescent current without affecting the bias condition.

The next portion of the circuit consists of a switch, S1, wired across the resistor R1. This switch is normally open, and is the equivalent of the trigger source for the latch, as we had in the previous SCR latch circuit. The circuit would normally be in the nonconducting state when we observe the monitor circuit consisting of LED (D1) wired across the latch output from pin #4. Resistor R4 (1 kohm) is the current limiter for D1. When power switch is turned on, the output of the latch (as monitored by D1) is off. Turn on S1 to simulate the feed action of the trigger voltage; the input to pin #1 is taken to a logic high level and the circuit is latched on, as shown by D1 lighting up. Now turn S1 to the off position and we can see that the latch remains in the off state even when the trigger source is removed. To reset the system, keep the switch S1 in the off position and turn power switch S2 to the off position. This circuit simulates the previous simpler SCR based latch circuit.

Construction Tips

There'll be a fair amount of components needed for this build so allow plenty of space around the 14-lead DIL IC socket used to house the MC 14001 NOR gate. Two of the quad two-input NOR gates are used in this design. Power is applied to pin #14 (located at the upper far left of the package) and the ground connection is made to pin #7 (located at the lower far right of the package). On the circuit schematic you'll see only one of the ICs (IC1a) with the power and ground connections marked, and none on the second IC (IC1b); this is because both ICs are contained in the same package. The NOR gates (we've chosen to use) are located on the lower half of the package so take care in making the connections as the components will be fairly close to each other. Ensure also that the cross-coupled links going from the output (pin #3) of IC1a to the input of IC1b (pin #5), and from output of IC1b (pin #4) to the input of IC1a (pin 32), are routed correctly. It's easy to make a wiring mistake here. Because of the switches (S1 and S2) needed for verifying the correct operation of this device, a project case is essential. The rest of the passive components R2 and C1 going to pin #1 of IC1a, and R3 and C2 going to pin #6 of IC1b, are straightforward connections and won't pose any problems. To minimize the possibility of circuit build errors, take note that these NOR gates do not face the same way in the package layout—that is, they are located "end to face," so be especially vigilant here. Mark up the on/off positions for the switches S1 and S2, as their positions will determine the verification of the state of the latch output.

Test Setup

Testing is straightforward and follows the same sequence for the previous SCR latch. Switch S1 is first set to the off position. Next turn on power switch S2. The latch is now in the equilibrium state and remains nonconducting (LED D1 is off), for as long as there is no trigger voltage (from switch S1). Next turn S1 to the on state. The latch will flip over to a conducting state with LED D1 turning on. With the trigger voltage removed (by turning off switch S1), the latch continues to remain on. To reset the latch, power switch S2 needs to be turned off. When power switch S2 is returned to the on state, the latch will once more be in the standby off state.

Parts List

Semiconductors
    IC1a: 1/4 MC 14001 CMOS quad two-input NOR gate
    IC1b: 1/4 MC 14001 CMOS quad two-input NOR gate
Resistors (all resistors are 5 percent, 1/4 W)
    R1: 1 Mohm
    R2: 100 kohm

R3: 100 kohm
R4: 1 kohm
R5: 4.7 kohm
Capacitors
(All nonpolarized capacitors disc ceramic)
(All electrolytic capacitors have a 25 V rating)
C1: 0.047 µF
C2: 0.047 µF
C3: 0.1 µF
C4: 100 µF
Additional materials
D1: LED
D2: LED
S1: single-pole, single-throw miniature switch
S2: single-pole, single-throw miniature switch
Power supply: nine-volt battery

Project 17: NOR Gate Metronome

Introduction

The versatility of the basic NOR gate is shown in this unusual metronome circuit. A metronome is essentially an ultra-slow-speed square-wave oscillator, running somewhere in the region of a few hundred BPM (beats per minute), used by musicians as a timing device. This circuit uses two NOR gates.

Circuit Description

The MC 14001 quad two-input NOR gate is configured here to act as a very-slow-speed square-wave oscillator. The schematic is shown in Figure 7-4.

Two NOR gates are used to form a square-wave oscillator configuration, and as we would expect, a resistor and capacitor combination controls the frequency of the generated output. IC1a (the first NOR gate) has the two inputs (pins #1 and #2) shorted together. A resistor R1 (100 kohm), potentiometer VR1 (50 kohm), and resistor R2 (2.7 kohm) form a series chain between the paralleled inputs (pins #1 and #2) and the output (pin #3). Eventually, VR1 will act as the variable control for the metronome speed, defined in BPM (beats per minute). IC1b (the second NOR gate) has its inputs (pins #5 and #6) similarly paralleled together. The output from IC1a is fed into the paralleled input of IC1b. The output of IC1b (pin #4) is returned back through capacitor C1 (10 µF) to the junction of R1 and VR1. Lowering the value for C1 will raise the frequency of the output waveform, but it's recommended you stay with the value given. Across the range of VR1's rotation, you'll see a metronome speed span of 30 to 220 BPM (or 0.5 Hz to 3.7 Hz). For improved stability of speed, a tantalum (rather than the usual electrolytic) capacitor is defined for C1. This is important for the metronome, where you're relying on the beats per minute

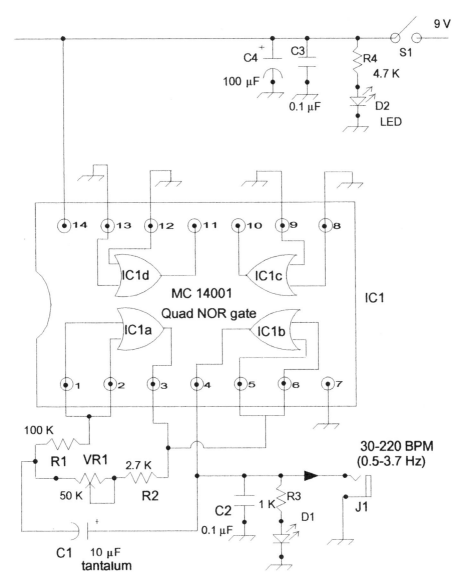

Figure 7-4 NOR Gate Metronome

remaining constant. The metronome output is taken from pin #4 of IC1b and goes straight into a LED monitor D1. Resistor R3 (1 kohm) is the current limiter for D1. A capacitor C2 (0.1 µF) to ground from pin #4 of IC1b stabilizes the output. The circuit runs immediately on switch on. Switch S1 applies nine volts to power the circuit. LED D2 and current-limiter resistor R4 (4.7 kohm) acts as the supply voltage indicator. Capacitors C3 (0.1 µF) and C4 (100 µF) stabilize the supply voltage against transients. Unused inputs (pins #8, #9, #12, #13) are grounded to prevent spurious responses. Output LED D1 flashes in sync with the positive-going square-wave pulses and can be used to pace your musical endeavors.

Construction Tips

There are quite a few cross-coupled components in this circuit, so take your time; work slowly to eliminate the chance of wiring errors. Link the inputs (pins #1 and #2) together first for IC1a. Similarly link the inputs (pins #5 and #6) together for IC1b. Add in the series resistor chain first, consisting of R1 (100 kohm), VR1 (50 kohm), and R2 (2.7 kohm), between IC1's inputs (pins #1 and #2) and the output (pin #3). Next add in the link from IC1b's inputs (pins #5 and #6) to IC1a's output (pin #3). The next connection is from IC1b's output (pin #4) via tantalum capacitor C1 (10 μF) to the junction of resistor R1 (100 kohm) and potentiometer VR1 (50 kohm). Note the polarity of C1 (10 μF); the positive end goes to IC1b's pin #4, and the negative end goes to the junction of R1 and VR1. The output is taken from IC1b's pin #4 to resistor R3 (1 kohm) and LED monitor D1. Finally, add in capacitor C2 (0.1 μF) across IC1b's output (pin #4) also.

Test Setup

Potentiometer VR1 is the BPM controller and can be calibrated by counting the number of beats per minute for various settings of VR1 and marking up the control knob. The markings can be approximate, as you're likely just to be adjusting VR1 to a convenient value that suits the music you're using.

Parts List

Semiconductor
    IC1: MC 14001 quad two-input NOR gate
Resistors (all resistors are 5 percent, 1/4 W)
    R1: 100 kohm
    R2: 2.7 kohm
    R3: 1 kohm
    R4: 4.7 kohm
Capacitors
(All nonpolarized capacitors disc ceramic)
(All electrolytic capacitors have a 25 V rating)
    C1: 10 μF (tantalum)
    C2: 0.1 μF
    C3: 0.1 μF
    C4: 100 μF
Additional materials
    VR1: 100 kohm potentiometer
    D1: LED

D2: LED
S1: single-pole, single-throw miniature switch
Power supply: nine-volt battery

Project 18: 74LS122 Monostable

Introduction

One of the perverse aspects of trying to use application note circuits (you know, those circuits you find in manufacturer's data books recommending how their devices could be used) is that they're never real circuit schematics. There are unmarked pins that have you wondering what to do with them—and what about those various input pins? What goes to them, and how are the interconnections made? That's an application note for you. It's only a suggested scheme. To make the circuit run requires usually a lot more thought and components. A real circuit, as you'd find in this book, has a description of every connection made to each and every terminal. There's no guesswork needed. So here we'll take a look at a very useful device called the monostable, from the point of view of converting the application note into a real circuit. The application note usually just gives you the required information for calculating the timing pulses. A monostable is a circuit that generates a variable pulse width (determined by a resistor and a capacitor) when triggered by a positive input voltage. For the specific device in question, we'll be looking at the 74LS122 TTL device running off a five-volt supply.

Circuit Description

The monostable circuit is shown in Figure 7-5.

Let's start with an oddity that's found in the application note and work to simplify it. The timing components for the 74LS122 monostable are simple enough, just a resistor R1 (100 kohm) and a capacitor C1 (100 µF). But here's the twist. The generated pulse width period (as per the application note) is given by:

$$t = 0.37 \times R1 \times C1$$

where t is in nanoseconds, R in kohms, and C in picofarads. Considering that we're usually more familiar with working with t in seconds and C in microfarads (R in kohms is fine), the stated units appear odd. So with a little math conversion we can make the calculation much more user friendly. Here are the two versions (to show they produce the same results).

Using the circuit values of R1 = 100 kohm and C1 = 100 µF, the value for pulse width is calculated as: t (in nanoseconds) = 0.37×100 (in kohm) $\times 100.10^6$ (in pF). Note the conversion of 100 µF to 100.10^6 pF. The calculated pulses width thus works out to be $3.7.10^9$ nanoseconds, or 3.7 seconds. Now use a simpler equation of mine for the same calculation: t (seconds) = $3.7 \times 10^{-4} \times R$

124 BEGINNING DIGITAL ELECTRONICS THROUGH PROJECTS

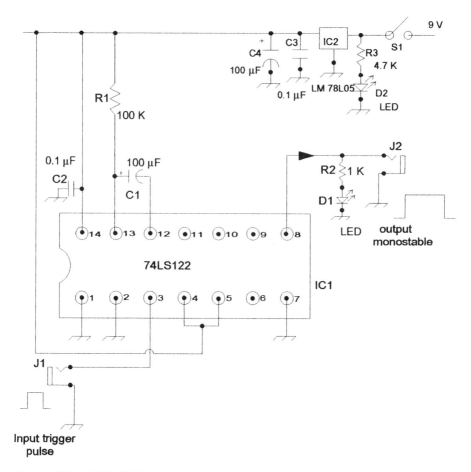

Figure 7-5 74LS122 Monostable

(in kohm) × C (in µF). Using the same component values as before for R1 (100 kohm) and C1 (100 µF), we get: t (seconds) = $3.7 \times 10^{-4} \times 100 \times 100 = 3.7$ seconds, which is the same as before, but this time it's a much more appropriate process of calculation!

The circuit itself is very simple—the critical timing components are just a resistor R1 (100 kohm) and a capacitor C1 (100 µF). The active device itself is IC1 (74LS122), a TTL IC running off 5 volts. Take a look at the circuit schematic. It's unusual in the sense that it is drawn showing the IC package outline rather than the symbol, as you'd see for an operational amplifier. Digital circuits are often shown this way; it depends on the particular device itself that's being used. IC1 is a 14-lead DIL device, with power applied to pin #14 (where it usually goes), and ground is taken from pin #7. Capacitor C2 (0.1 µF) decouples the power pin (#14). The timing component R1 (100 kohm) is connected from pin #13 to Vcc, and C1 (100 µF) is coupled across pin #13 and #12. The positive end of C1 goes to pin #13. The output is taken from pin #8 and terminates in a LED (D1) and current-limiter resistor R2 (1 kohm). The trigger source is a positive-going pulse (narrow width) fed to pin #3. Timing of the output pulse starts on the positive-going transition of the input trigger. Pins #1 and #2 are grounded. Pins #4 and #5 go to Vcc. The supply voltage has to be five volts, and regulator IC2 (78L05) is used again here. Power switch, S1, is the on/off controller. Supply voltage monitoring is via LED (D2) and current-limiter resistor R3 (4.7 kohm). Capacitors C3 (0.1 µF) and C4 (100 µF) enhance stability.

Construction Tips

Use a 14-lead DIL IC socket for housing the 74LS122. A feed input socket, J1 (1/8-inch miniature jack socket) will do nicely to couple in the trigger pulse. Do the same for the output, using J2 (1/8-inch miniature jack socket) to couple an oscilloscope. There are only a few components needed, so the risk of error is small. Mount the mechanical devices on a small project case.

Test Setup

You're going to need a positive-going single-shot TTL trigger pulse to initiate this circuit, as the monostable requires a clean positive transition to engage. Any short trigger-pulse width will do—that is, less than the designed width of the monostable output. As the output pulse width is deliberately long to facilitate viewing (3.7 seconds), you've got plenty of flexibility to use anything less than that. A 1-millisecond pulse width is a good starting point. You can use a function generator to set up a pulse train that produces a low-duty-cycle output, with a repetition rate in excess of 3.7 seconds.

That way, you get the following sequence. The function generator produces a short-pulse-width trigger. On the trigger's rising positive edge, the monostable triggers, and a 3.7 second pulse width output is formed. LED D1 turns on. The function generator output repeats, and the sequence cycles again. Let's say you set the trigger signal repetition rate to be five seconds (that's a frequency of 1/5 seconds = 0.2 Hz); then you can see the operation is effectively in slow motion. If the sequence is too slow, lower the value for capacitor C1, and the monostable pulse width decreases. The trigger repetition rate can be sped up accordingly. If you have an oscilloscope, use it to monitor the input trigger on channel 1 and the monostable on channel 2. This will give you a nice visual display of the monostable's operating sequence.

Parts List

Semiconductor
　　IC1: 74LS122 TTL monostable

Resistors (all resistors are 5 percent, 1/4 W)
    R1: 100 kohm
    R2: 1 kohm
    R3: 4.7 kohm
Capacitors
(All nonpolarized capacitors disc ceramic)
(All electrolytic capacitors have a 25 V rating)
    C1: 100 µF
    C2: 0.1 µF
    C3: 0.1 µF
    C4: 100 µF
Additional materials
    D1: LED
    D2: LED
    S1: single-pole, single-throw miniature switch
    Power supply: nine-volt battery

Project 19: 74LS75 Quad Latch

Introduction

This circuit is still based on a single digital IC, but it's getting more complex with the inclusion of more switches and monitoring LEDs. Often in digital circuit applications we can benefit from an indication of when a digital event has taken place. Rather than monitoring a circuit, trying to see when an event has occurred, you can have the latch do the monitoring remotely for you (we saw this earlier in the introductory latch circuits). The 74LS75 latch transfers whatever data we supply to it to a storage area, where it will stay as long as power is maintained to the circuit. This circuit is a complete four-channel latch, and it shows the versatility that can be effected with using just one IC.

Circuit Description

IC1 is a dedicated TTL 74LS75 device called a latch. This is a device able to transfer whatever data (logic high or low) is at an input to its unique output storage location. Data is transferred from the inputs to the outputs (each of the input/output pairs being paired up) through the controlling action of an enable pin. Think of the enable pin as the traffic controller for the routing action. Data transitions from an input to its paired output are made when the enable pin goes high. The schematic is shown in Figure 7-6.

A four-bit latch such as the 74LS75 matches perfectly to digital applications where you would be using four-bit words and would want to store each of the individual bits. The latch outputs are bistables, meaning they will remain permanently in whatever logic state they've been transferred to. Data for latch

Versatile Digital ICs: Projects 14–20 **127**

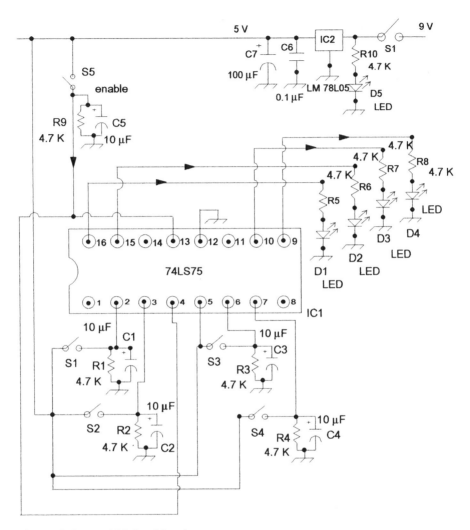

Figure 7-6 74LS75 Quad Latch

#1 is fed to pin #2. Latch #1's output is taken from pin #16. LED D1 and resistor R5 (4.7 kohm) act as monitors for latch #1's output state. In this particular setup, we have a switch, S1, applying a positive voltage (derived from the Vcc positive supply rail) to latch #1's input. When the switch (S1) is open circuit, a resistor R1 (1 kohm) and capacitor C1 (10µF) combination pulls the input down to a logic low. The inclusion of R1 and C1 is important. Without R1 and C1, the input would float (that is, you could never determine its logic state precisely), leading to an undesirable unknown logic state at the output. The positive end of C1 is coupled directly to the input. Data is transferred from the input to the output only when the enable (pins #4 and #13) is taken high. In this circuit, the enable is selectively set to a logic high through switch S5, which makes the connection to Vcc. In the off state of S5, a resistor R9 (1 kohm) and capacitor C5 (10 µF) combination keeps the enable low.

The latching action is as follows. As long as the enable is kept high, the latch output will follow the input. When the enable goes low, the previous logic state of the output is retained, regardless of the subsequent state of the input. Only when the enable is taken high again will the output track the input. The rest of the quad latches are configured exactly as the first. The connections are:

Latch #1:	input/pin #2	output/pin #16
Latch #2:	input/pin #3	output/pin #15
Latch #3:	input/pin #6	output/pin #10
Latch #4:	input/pin #7	output/pin #9

Power is applied to pin #5 located along the lower edge of the IC. The ground connection is made to pin #12, located along the upper edge of the IC. These power/ground connections are not where you'd normally expect them in a digital device, so take care when making the connections. The five-volt power requirement is met through a regulator IC2 (LM78L05) with power switch S6. LED D5 and limiter resistor R10 (4.7 kohm) act as a power-on indicator. Capacitors C6 (0.1 µF) and C7 (100 µF), on the output side of IC2, stabilize the supply voltage.

Construction Tips

IC1 requires a 16-lead DIL socket. There are six switches in all for this build and five LEDs, so a larger project case than usual is called for. Arrange the switches to match the output LEDs they correspond to, in order to facilitate monitoring of the latching function. Mark up the high/low logic positions for the input switches and enable switch on the front panel; that way you can identify their polarities without confusion.

Test Setup

This circuit is entirely self-contained, requiring no external stimulus or output monitor. When the enable switch (S5) is taken high, whatever logic state is given to the input will be transferred immediately to the output. When the enable is taken low, the output locks at the previous logic state it was set to, illustrating the latching action.

Parts List

Semiconductors
  IC1: 74LS75 TTL quad latch
  IC2: LM78L05 regulator

Resistors (all resistors are 5 percent, 1/4 W)
- R1: 4.7 kohm
- R2: 4.7 kohm
- R3: 4.7 kohm
- R4: 4.7 kohm
- R5: 4.7 kohm
- R6: 4.7 kohm
- R7: 4.7 kohm
- R8: 4.7 kohm
- R9: 4.7 kohm
- R10: 4.7 kohm

Capacitors
(All nonpolarized capacitors disc ceramic)
(All electrolytic capacitors have a 25 V rating)
- C1: 10 μF
- C2: 10 μF
- C3: 10 μF
- C4: 10 μF
- C5: 10 μF
- C6: 0.1 μF
- C7: 100 μF

Additional materials
- D1: LED
- D2: LED
- D3: LED
- D4: LED
- D5: LED
- S1: single-pole, single-throw miniature switch
- S2: single-pole, single-throw miniature switch
- S3: single-pole, single-throw miniature switch
- S4: single-pole, single-throw miniature switch
- S5: single-pole, single-throw miniature switch
- S6: single-pole, single-throw miniature switch
- Power supply: nine-volt battery

Project 20: CD4072 Quad Switcher

Introduction

This circuit can be used to form the basis of a four-site detector, such as a flood detector in your basement. Any one of the trigger sensors that is set off by the presence of unwanted water can initiate an alarm circuit. The quad input OR gate responds if any of the detectors are triggered. The circuit's mechanical switches are used to represent the sensors, which can be easily implemented by any one of a number of trip sensors. The digital logic device is the core mechanism.

Circuit Description

IC1 is a CD4072 quad input OR gate shown in Figure 7-7. The OR gate's output will go into a logic high state when any one of the inputs go high. At the initial start-up state, all the four inputs are taken low until an event sets one of the four inputs high. The OR gate inputs are located at pins #2, #3, #4, and #5. Pull-down resistors keep the inputs locked at a logic low level. For input #1, resistor R1 (1 kohm) and capacitor C1 (10µF) are the pull-down network. The low value resistor R1 keeps the input at a logic low. Capacitor C1 guards the input against unwanted transients falsely triggering the device. Switch S1 couples the input to a logic high (five-volt Vcc). Each of the inputs has a similar arrangement, shown as:

Input #1 (pin #2):   Resistor R1 (1 kohm), Capacitor C1 (10µF), Switch S1
Input #2 (pin #3):   Resistor R2 (1 kohm), Capacitor C2 (10µF), Switch S2
Input #3 (pin #4):   Resistor R3 (1 kohm), Capacitor C3 (10µF), Switch S3
Input #4 (pin #5):   Resistor R4 (1 kohm), Capacitor C4 (10µF), Switch S4

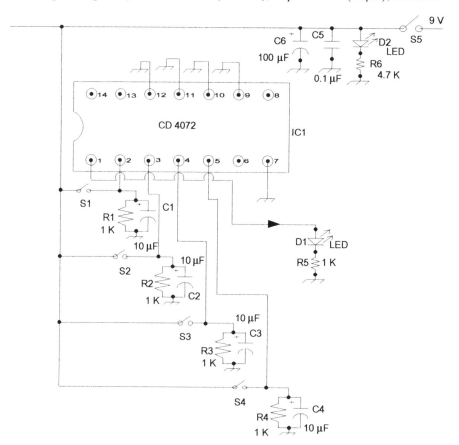

Figure 7-7 CD4072 Quad Switcher

The output is taken from pin #1, with an LED (D1) acting as the logic state monitor. Resistor R5 (1 kohm) provides current limiting for D1. As this is a dual device, there is a second identical quad OR gate in the package. We are not using this second set, so the unused inputs are grounded. The pins to be grounded are #9, #10, #11, and #12. Power is applied to pin #14, and ground is taken to pin #7. As this is a CMOS device, the supply voltage can be anywhere between 3 and 18 volts, so a nine-volt supply is fine. Power switch S5 is located at the voltage-feed side. The application of power is indicated by an LED (D2) and current-limiter resistor R6 (4.7 kohm). Capacitors C5 (0.1) and C6 (100 µF) stabilize the circuit against power-supply transients.

Construction Tips

IC1 requires a 14-lead DIL socket to house the CD4072. This is a CMOS device, so take the usual antistatic handling precautions. A project case is necessary because of the high switch count in this circuit (there are five of them). All of the used inputs for OR gate #1 are located next to each other (pins #2, #3, #4, and #5) and require the same resistor/capacitor combinations, so component location is not convoluted. Power and ground connections are at the conventional extreme upper left and extreme lower right positions.

Test Setup

To verify the circuit operation, first set all of the input switches (S1, S2, S3, S4) to the open-circuit position. Next turn on the power through switch S5. The output from pin #1 will remain off, as indicated by the LED D1 staying unlit. Engage switch S1 to the high position, and you'll see LED D1 turn on. As this is an OR circuit, any one of the inputs going high will turn the output on. To verify the rest of the OR gates, switch S1 to the off position and then sequence the rest of the switches (S2, S3, S4) individually to the high state. With each transition, the output monitor LED (D1) will turn on.

Parts List

Semiconductor
    IC1: CD4072 quad input OR gate
Resistors (all resistors are 5 percent, 1/4 W)
    R1: 1 kohm
    R2: 1 kohm
    R3: 1 kohm
    R4: 1 kohm
    R5: 1 kohm
    R6: 4.7 kohm

Capacitors
(All nonpolarized capacitors disc ceramic)
(All electrolytic capacitors have a 25 V rating)
    C1: 10 μF
    C2: 10 μF
    C3: 10 μF
    C4: 10 μF
    C5: 0.1 μF
    C6: 100 μF
Additional materials
    D1: LED
    D2: LED
    S1: single-pole, single-throw miniature switch
    S2: single-pole, single-throw miniature switch
    S3: single-pole, single-throw miniature switch
    S4: single-pole, single-throw miniature switch
    S5: single-pole, single-throw miniature switch
    Power supply: nine-volt battery

CHAPTER **8**

Digital Support Circuits: Projects 21-26

The previous chapter introduced you to a range of commonly encountered digital circuits. This chapter offers a collection of supporting circuits that expand the range of the basic digital device. Real circuits often require digital circuits to be interfaced into a variety of loads, which are more often than not low-impedance loads. The first project for this chapter, therefore, features a simple but effective high-current TTL driver that can be used to increase the basic current-driving capability of any TTL logic device. The subsequent circuits will show their admirable usefulness as your experimentation with digital circuits extends into the practicalities of real, working (as opposed to textbook) circuits.

Project 21: TTL Relay Driver

Introduction

Digital circuits produce nice, clean pulses, and it's a very easy matter to interface a chain of such circuits together to produce predictable digital results. For example, if we take the two inputs of a dual input AND gate to a logic high level, the output goes predictable high. That's the benefit of working with digital circuits; for a particular logic device, the output logic state is very clearly determined by how the state logic inputs are set up. When it comes to real applications however, which means actually putting digital circuits to work in a real project, we're usually going to need a little help. The difference between a textbook logic circuit that behaves as predicted and a "real, working" circuit that is generally expected to drive a low-impedance load or interface into a follow-on circuit is quite significant. Most of the ancillary circuits described in this chapter will show you how the interfacing can be done, thereby enhancing the base-level capability of stock digital circuits. Most of the time, when you're working with straight digital logic ICs the logic output state can be easily verified with a simple monitor, usually just an LED voltage-level indicator (if the pulse-repetition

133

rate is low). If you want more detail (for faster pulse-repetition rates), the oscilloscope is preferred.

Digital circuits inherently are not designed to provide any significant current into a low-impedance load. When working with analog circuits, such as ac audio amplifiers to feed into low-impedance loads, there is a really nice, ready solution. This is the versatile LM 386 audio power-amplifier integrated circuit, which is used almost universally for driving low-impedance loads, such as speakers. Life would be simpler if there was an easy-to-use digital equivalent. Since this is not the case (digital circuits are dc and not ac driven), we have to resort instead to other discrete techniques. Most commonly, we would be looking to find a dc high-current driver, i.e., one that can be interfaceable to TTL logic level signals. One of the simplest ways to fill this requirement would be to use a simple single transistor current booster that is capable of driving a low resistance load such as a relay. The use of a low-voltage relay (which can be conveniently driven from a nine-volt supply) is an ideal means for providing an interface driver for high-current loads. Typically a relay's contacts can support load currents that are even in the "amps" range with voltage handling maximums even up to the 110 V line voltage. So that's the circuit we have described here as the first project of this chapter; the TTL relay driver.

Circuit Description

TTL circuits by themselves are incapable of directly driving any load that draws significant current; but just add a simple relay driver and the capability difference is remarkable! Using nothing more than a single readily available general purpose NPN transistor, we can come up with a really effective TTL relay driver circuit as shown in Figure 8-1.

Q1 is a 2N2222 NPN transistor that is the driving heart of the circuit. There's nothing special about the choice of this transistor and many other general purpose types can be used as alternates. But, the 2N2222 transistor is one I've often used for lots of projects, and it's a very dependable device. When you find a device that is reliable, you generally like to stick with it. The choice of transistor though has to be an NPN type in order to be driven from a positive supply voltage, as digital circuits run off a positive supply.

A transistor (if this is your first acquaintance with them) is a three-terminal device with a collector, base and emitter terminal. Generally it is required that the correct dc bias conditions are set up, before the device can perform any useful function. That's one of the fundamental differences between working with transistors and integrated circuits. For example, consider the simple case of the design of an ac audio amplifier circuit. The stage gain of a transistor based amplifier is going to influence by not only the values of the biasing components used, but also the transistor's characteristics—change the transistor to a similar type and the gain will change. An integrated circuit amplifier on the other hand has a much simpler way to determine the gain; it's governed simply by the ratio of two resistors and the gain is essentially

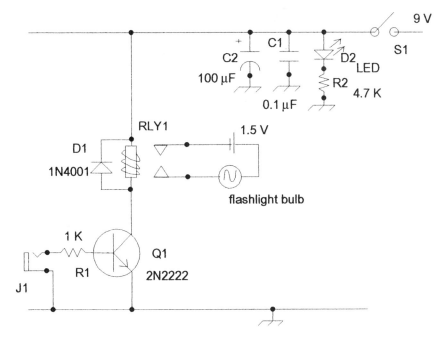

Figure 8-1 TTL Relay Driver

independent of the IC itself—change the IC to a similar type and the gain remains the same. In the age of integrated circuits, therefore, the lowly transistor might be dismissed as just a device of no particular significance, but this is not true (as you'll see throughout this chapter) as it's the cleverness of the circuit design that counts and not the complexity of the device itself.

This TTL relay driver is designed to take an input feed from a positive digital TTL source. In the schematic the feed source is shown as being coupled directly into an input resistor R1 (1 kohm). The purpose of this resistor is to limit the current flowing into the transistor's base terminal. When the input voltage (that is applied to the base terminal) exceeds a certain positive threshold value, the transistor will turn on, i.e., conduct, allowing a relatively higher current to flow through the output collector terminal. The load (in this case the relay) is placed in series with the collector terminal. The absolute maximum rating for the collector current flow will be dictated by the choice of transistor used. Typically, for the 2N2222 you can expect to handle around 600 milliamps maximum. For this application we can use a low-voltage relay, such as a nine-volt mini-relay type RLY1, which has a single-pole, single-throw (SPST) set of output contacts.

A relay is characterized by the electrical parameters of two separate sections: the primary drive coil and the secondary load contacts. The relay's primary drive coil has a low impedance, typically 500 ohm (for the specified type of device in the parts list). The resultant current drawn through the primary coil is quite low, only 18 milliamps in value (9 volts/500 ohm = 0.018 amps, from Ohm's Law). The output set of contacts is rated in terms of its drive capability. This relay is specified as having a two-amp drive capability, which means you can drive really high-power loads. Commonly with any circuit involving a relay, you will see a protection diode placed across the relay; in the schematic, this protection diode is shown as D1 (1N4001). The purpose of the diode is to protect transistor Q1 from any inductive surges that occur across the relay coil. These surges occur when the coil is pulsed on and off as a consequence of a pulse train being applied to the input. To complete the electrical connections to the circuit, the emitter terminal is grounded in order to form the return path for the signal.

The supply feed for the circuit comes from a nine-volt battery, that is, fed in from a power switch, S1. LED D2 and resistor R2 (4.7 kohm) function as the on/off monitor for the supply voltage. Capacitors C1 (0.1 µF) and C2 (100 µF) are included to stabilize the supply voltage.

Construction Tips

Most of the space on the assembly board will be taken up by the type of relay (the largest board component) you decide to use. Principally, you need to select a type of relay (if you can't locate the suggested type) that's activated by a nine-volt supply and has a primary coil current requirement that can be supplied from that battery. There are many types of relays available, and this is the main criterion when selecting a relay. Other than that, the relay's secondary contacts (which are going to drive the load) can be any type, as you're basically going to need an on/off switching arrangement. It's advisable to mount the relay onto the board and identify the primary and secondary leads. The transistor can be worked on next. The transistor's leads (emitter, base, collector) need to be first identified and then gently separated so the device can be inserted into the assembly board.

The following orientation is one I use when aligning the leads: the emitter terminal faces "south," the base terminal faces "west," and the collector terminal faces "north." This arrangement makes it easier to add the other components so the final circuit closely resembles the circuit schematic. Checking the circuit is much simplified, as the component placement will closely match the layout on the schematic. Since there are so few components involved in this project, the chances of assembly error are slim. The protection diode, D1, must be located with the orientation as shown. Inadvertent reversal of D1 will cause the diode to be forward biased, essentially shorting out the relay coil and causing the circuit to malfunction.

Test Setup

For the test setup you'll need a TTL pulse source that's either a single-shot (one pulse) or has a low frequency-repetition rate, as the relay (being a mechanical device) can only switch on relatively slowly. Any high-current source and load can be connected across the relay's secondary contacts. Note the relay also provides total electrical isolation between the primary (coil) and the secondary (contacts) circuit. As a circuit to test out the high-current drive capability (of the relay), you could simply couple up a flashlight bulb to a separate 1.5-volt battery and use the relay contacts as the intermediate switch. The flashlight bulb draws a large current because of the very low resistance of the filament. When the circuit is driven by a TTL pulse source, you'll see the flashlight bulb light up each time the input pulse goes high. Without the relay driver interface the TTL pulse does not have the drive capability to power up the flashlight bulb directly.

Parts List

Semiconductor
   Q1: 2N2222 NPN general purpose transistor
Resistors (all resistors are 5 percent, 1/4 W)
   R1: 1 kohm
   R2: 4.7 kohm
Capacitors
(All nonpolarized capacitors disc ceramic)
(All electrolytic capacitors have a 25 V rating)
   C1: 0.1 µF
   C2: 100 µF
Additional materials
   RLY1: SPST mini-relay (e.g., Radio Shack 275-005)
   D1: 1N4001
   D2: LED
   S1: single-pole, single-throw miniature switch
   Power supply: nine-volt battery

Project 22: High-Power Transistor Driver

Introduction

If you've got a really huge amount of current (and we're talking about a lots of amps here!) to sink into a load, try the ultimate in single-device power transistors, the macho 2N3055 metal-can TO-3 NPN driver! This is hefty, Rambo-sized transistor (it's physically quite a large and heavy transistor, nothing at all like the tiny 2N2222 small-signal device used in the previous circuit). The 2N3055 has a collector current drive capability of 15 amps! When you translate this current capability into a power rating, this is a massive power-handling capability of 115 watts, if we were to use a nine-volt voltage source. A nine-volt battery could not, of course, supply anywhere near this type of current, but if you had a power supply (one that was fed from the line voltage and transformer stepped down to nine volts) capable of supplying this amount of current, this enormous power is achievable. With that amount of

138 BEGINNING DIGITAL ELECTRONICS THROUGH PROJECTS

drive capability, the 2N3055 is practically indestructible! In practice the input drive voltage to the 2N3055 would need to be very high, which would necessitate having several stages of significant pre-amplification. Nevertheless, without having to run the 2N3055 to its limit, we can still show its extreme usefulness as a power driver in this project schematic.

Circuit Description

In the circuit schematic Q1 is a NPN 2N3055 power transistor configured in the common-emitter amplifying mode, with the load placed in series with the collector terminal. The schematic is shown in Figure 8-2.

There are two resistors, R1 (4.7 kohm) and R2 (470 ohm), which are used to provide a dc bias to the base terminal. Capacitor C1 (4.7 µF) is the ac coupler for an ac signal. This large-value capacitor is an electrolytic type and requires a polarity orientation with the capacitor's positive terminal going to the base terminal. The input feed signal could be either a sine, triangle, or square wave. Even with using a square-wave input, which normally makes a transition from the zero volts baseline to five volts (for a TTL signal), the input capacitor, C1, will remove the dc component and produce a signal that is locked to the zero volts line but moves equally in the positive and negative directions (essentially ±2.5 volts, if the feed was a five-volt TTL).

In this circuit the load goes directly in the collector circuit. Because of the incredible current-driving capability of the 2N3055, there is no relay needed, as there would be if using a less powerful device (as in the previous

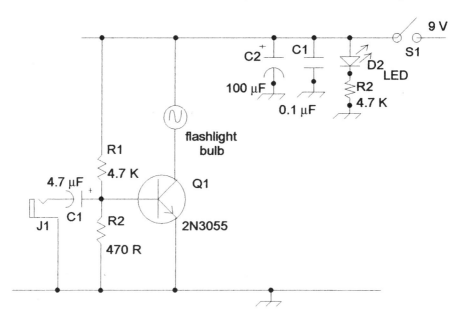

Figure 8-2 High-Power Transistor Driver

circuit). Because of the direct drive connection, you can switch Q1 at a much faster rate (you are not limited by the slow mechanical movement of the relay contacts opening and closing). The current flowing through the collector is essentially determined by the value of the load resistance. So for an example load resistance of, say, 10 ohms, the current is going to be nine-volt/10 ohm = 0.9 amp. Q1's emitter is grounded for this configuration to complete the signal path.

For intermittent use, Q1 can be used as is, but for continuous high-current use, heat sinking of the device to a metal chassis is recommended, in order to dissipate the heat being generated. The top of the device will quickly get hot, depending upon the current being drawn. The collector terminal is electrically connected to the case of the 2N3055, so take care when heat sinking the unit. Use a mica (insulating) washer to obtain electrical isolation. You can get a kit of mechanical parts for heat sinking the 2N3055 from component stores. Note that to drive the device to its limit, a succession of intermediate driver stages are necessary to bring the input signal to a high enough level. In this example, the collector load is shown as a regular flashlight filament bulb (choose a value rated to withstand nine volts), which will draw a high current because of its low resistance coil (filament) construction.

Power is fed through switch S1 from a nine-volt battery. LED D2 and resistor R2 (4.7 kohm) make up the supply monitor. Capacitors C1 (0.1 μF) and C2 (100 μF) smooth out the supply.

Construction Tips

The 2N3055 is one massive device! It's going to take up the majority of the assembly board area, dwarfing the rest of the resistors and capacitors. The electrical connection to the collector terminal needs to be done through a solder tag anchored via a nut/bolt arrangement to the device's body (there are two holes already provided for this purpose). Connections to the base and emitter are thick-solder lead terminations emerging from the underside of the device.

It's best to bring out three flying connections for the collector, base, and emitter terminals, since the 2N3055 is meant to be mounted on heat sink, in a real application. Cluster these three terminals in a configuration that corresponds to the schematic, that is, emitter facing "south," base facing "west," and collector facing "north." The signal input feed comes in via an electrolytic capacitor, C1, whose positive terminal goes to the base terminal. The bias resistors, R1 and R2, extend from base terminal to the Vcc and ground rails respectively. This is a straightforward routing and shouldn't pose any problems.

Don't forget the return signal path from the emitter to ground. For the load setup you can use a flashlight bulb as a demo. The best way to make the electrical connections is via a flashlight bulb socket. These have convenient screw terminations to which you can easily bring out a further two flying leads.

It doesn't matter which way the flashlight build leads are connected. Make sure when selecting the flashlight bulb, the rated value is sufficient to withstand nine volts. If you use too low a value, such as 1.5 volts, it's likely that when you connect up all you'll see is a brilliant white flash as the filament terminates in a blaze of glory. Either a visual examination of the broken filament or a continuity check will verify the destroyed bulb.

Test Setup

A function generator is the best input test source, as there's plenty of voltage available and the frequency can be slowed down to a viewable speed for the output. Switch on the power to the 2N3055 before turning on the function generator and couple in a low-frequency (a few tens of hertz is a good start) signal so you can see the switching action occurring. As you increase the signal input (use a triangle wave to start, as it rises and falls linearly), the voltage at which the device turns on can be easily seen. The flashlight bulb will turn on to signify that the collector current is flowing. Once you've determined the threshold switching levels, increase the signal frequency to see the effect on the output load. Eventually, the bulb will stay continually on, as the frequency is increased beyond the filament's ability to track the rapid changes. As an interesting experiment, switch over to a square-wave input, starting with a 50 percent duty cycle, taking the frequency to the point at which the build appears continually lit. If you now reduce the duty cycle, you'll see the bulb dim. This in effect is the technique used by a domestic light dimmer to dim your house lights.

Parts List

Semiconductor
    Q1: 2N3055 NPN high-power transistor
Resistors (all resistors are 5 percent, 1/4 W)
    R1: 4.7 kohm
    R2: 470 kohm
    R3: 4.7 kohm
Capacitors
(All nonpolarized capacitors disc ceramic)
(All electrolytic capacitors have a 25 V rating)
    C1: 4.7 µF
    C2: 0.1 µF
    C3: 100 µF
Additional materials
    D1: LED
    S1: single-pole, single-throw miniature switch
    Power supply: nine-volt battery

Project 23: Power Output FET

Introduction

The power field effect transistor (FET) is a very efficient solid-state switcher device that is ideally suited to acting as the final output stage for low-impedance loads. It takes up less space (physically) than the previous 2N3055 power transistor. The power output FET project is an upgraded version of the previous circuit.

Circuit Description

Any low-impedance load needs an appropriate power output stage that can supply both the voltage and current requirements. The power output stage described here is supplied by an FET. Q1 (IRF 610) is the power output FET, which runs conveniently off a five-volt (TTL) positive supply in order to make it compatible with handling a TTL digital signal. The FET is a three-terminal device, with source, gate, and drain electrodes, which are similar to the emitter, base, and collector terminals, respectively, for a transistor. The circuit in Figure

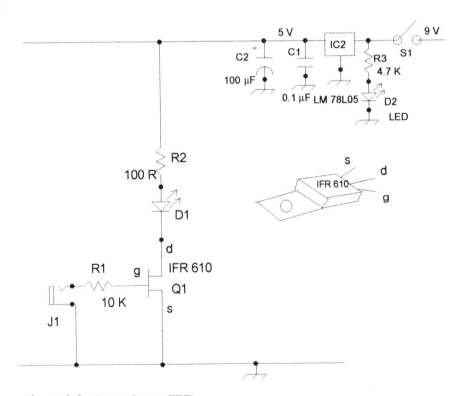

Figure 8-3 Power Output FET

8-3 is exactly the same as the transistor driver circuit described earlier. The input feed to the FET shown here is a dc source (there are no coupling capacitors used). The input resistor R1 (10 kohm) is a current limiter that couples the dc signal into the gate terminal of Q1. The output feed exits from the drain terminal. The circuit shows an LED (D1) with a low-value current-limiter resistor R2 (100 ohm) acting as the load. With this value for R1 (100 ohm), the LED current is 9 volts/100 ohm = 90 milliamps, which is higher than you would normally see in circuits where the LED is used as a voltage monitor. When a positive-going input TTL pulse is present, the FET turns on (conducting when a certain threshold voltage is exceeded), causing a current to flow through the drain terminal and so turning LED D1 on. In order to complete the signal return path, the source terminal of Q1 is grounded.

The five-volt power requirement is supplied through IC2 (LM78L05 regulator), which converts a nine-volt battery feed into a stable five-volt supply. Switch S1 is the power enabler, with LED D2 and limiter resistor R3 (4.7 kohm) being the on/off monitor. Capacitors C1 (0.1 µF) and C2 (100 µF), located on the output side of the five-volt regulator (IC2), stabilize the supply voltage. For convenience of test, an input jack socket J1 is added to the circuit.

Construction Tips

For the FET's connection, begin by identifying the terminals for the FET and position these as: source terminal facing "south," gate terminal facing "west," and drain terminal facing "north." The connections can then line up with the way the schematic is drawn up. The FET load LED can be mounted directly on the assembly board.

Test Setup

Apply power to the power output FET and couple in a regular TTL signal (e.g., from a function generator) via resistor R1. With an oscilloscope applied at the input to R1, you can monitor the signal input. Using a slow pulse-repetition rate so you can see the transition of both the input signal (on the oscilloscope) and output signal via the LED. The LED monitor in Q1's drain terminal will flash on and off, in sync with the input TTL pulse. The calculated 90-milliamp current flowing through LED D1, although significantly more we'd usually use, is still less than the current-driving capacity of Q1, which is around three amps. However, since the particular five-volt regulator (IC2) used here has only a current supply capacity of around 100 milliamps, you can't get much more current supplied. To increase the supply current (in order to get more load drive capability), you only need to substitute a higher-current-capacity five-volt regulator. The real purpose of describing this circuit, though, is to give you feel for the actual component values to use in a real circuit. The schematic component values remain the same if you're going to convert to a higher-capability five-volt regulator.

Parts List

Semiconductors
    Q1: IRF 610 power FET
    IC1: LM78L05 regulator
Resistors (all resistors are 5 percent, 1/4 W)
    R1: 10 kohm
    R2: 100 ohm
    R3: 4.7 kohm
Capacitors
(All nonpolarized capacitors disc ceramic)
(All electrolytic capacitors have a 25 V rating)
    C1: 0.1 µF
    C2: 100 µF
Additional materials
    D1: LED
    D2: LED
    J1: 1/8-inch miniature jack socket
    S1: single-pole, single-throw miniature switch
    Power supply: nine-volt battery

Project 24: TTL Driver Buffer

Introduction

Digital TTL logic signals are low-level signals that interface admirably to a chain of other digital circuits. However, you often find when working with real circuits that there's a final stage to which a digital circuit needs to be connected. If this final stage is a power output circuit, a form of buffer circuit is recommended in order to ensure that the drive signal (to the output stage) is not only adequate in maintained voltage but also consistent, regardless of the type of originating digital signal. The TTL driver buffer ensures this requirement is met; it resembles in essence the function provided by an emitter-follower circuit (that's often used as a buffer stage). The TTL driver buffer circuit described here is actually a dc threshold detector circuit, which will take a standard low-level TTL signal and convert it into a high drive–level signal that is more than capable of supporting any power output stage. This output drive signal is constant and predictable, with a drive (current) capability to match any power output stage (such as the previously described power FET circuit).

Circuit Description

As with all manner of power output stages, there is usually a specific level of input signal level needed for the power output stage to run efficiently—hence the call for driver stages. If you examine any typical audio power amplifier circuit, you'll see a series of driver stages preceding the final output stage. That's the principle found in any series chain of amplifier design; a small weak starting signal is gradually amplified stage by stage, first with voltage gain, and later with current gain capability.

The TTL driver buffer uses one-quarter of a quad-operational amplifier, IC1 (LM 324), that is configured to function as an dc threshold detector. The circuit is shown in Figure 8-4.

Two resistors, R1 (100 kohm) and R2 (10 kohm), form a simple potential divider network. The free end of R1 is connected to the five-volt Vcc rail and the free end of R2 connected to ground. The junction of R1 and R2 is coupled to the op amp's inverting terminal (pin #2), thus setting up the reference

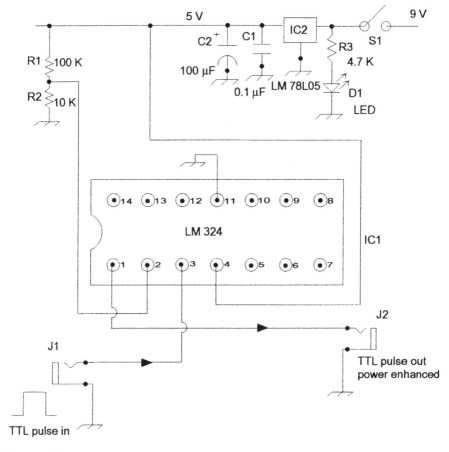

Figure 8-4 TTL Driver Buffer

voltage. The TTL signal input is coupled into the noninverting terminal (pin #3) of IC1. When the input signal exceeds the threshold value (as determined by the value of R1 and R2), the output (taken from pin #1) will switch from a low to high logic state. Since the TTL input signal is applied to the noninverting terminal, the boosted output signal will be in phase with the input signal. When the TTL input exceeds the trigger threshold, the output switches from a low to high state. This signal is in phase with the input but has a much higher current drive capability. Both the TTL input and dc output are directly coupled, through 1/8-inch miniature jack sockets J1 and J2, respectively, in order to assist with the routing of signals into and out of the TTL driver buffer. Power is fed in from a nine-volt battery via switch S1. LED D1 and resistor R3 (4.7 kohm) make up the supply monitor. A regulator IC2 (LM 78L05) produces the required five-volt supply for IC1. Capacitors C1 (0.1 µF) and C2 (100 µF) provide smoothing.

Construction Tips

It is recommended that a 14-lead DIL socket be used to mount IC1. There are only two resistors (R1 and R2) specific to the LM 324 itself, and these are soldered in close proximity to pin #2 of IC1. The rest of the components (LED D1, resistor R3, and capacitors C1 and C2) are the conventional ones clustered around the LM78L05 regulator IC. Since this circuit requires the connection of an input signal and an output signal, the jack sockets (J1 and J2) should be mounted on a project case. The power switch, S1, and LED D1 should also be mounted on the case. Label the jack sockets appropriately ("input," "output"), as they are identical. Conventionally, the input jack socket is mounted on the left side of the front panel and the output jack socket on the right-hand side.

Test Setup

A function generator and oscilloscope combination will allow you to monitor the switching action of the dc op-amp's threshold-detection action. Select a triangle waveform that is TTL compatible (making a zero to five-volt excursion), with a low repetition rate. Monitor the input signal on channel 1 on the oscilloscope. The output waveform from the TTL driver buffer can be monitored on channel 2 of the oscilloscope. Power up the circuit first, then turn on the function generator. You should see at once that the threshold voltage of about 0.5 volts is exceeded; the output will switch from a low (zero-volt) to high (five-volt) state. The threshold voltage is determined by: $V_{threshold} = [R1/(R1 + R2)] \times Vin = [10/110] \times 5 = 0.45$ volts. This equation derives from the basic potential-divider circuit.

Once you've verified that the circuit is operating correctly, you can change the input signal to a TTL signal, and from the two oscilloscope traces see that the signals are in phase—that is, both the signals go high and low together. If you have the previous circuit (power FET) constructed, this circuit can be coupled into the TTL driver buffer to verify that both circuits work together correctly.

Parts List

Semiconductors
    IC1: LM 324 quad operational amplifier
    IC2: LM78L05 regulator
Resistors (all resistors are 5 percent, 1/4 W)
    R1: 100 kohm
    R2: 10 kohm
    R3: 4.7 kohm
Capacitors
(All nonpolarized capacitors disc ceramic)
(All electrolytic capacitors have a 25 V rating)
    C1: 0.1 µF
    C2: 100 µF
Additional materials
    D1: LED
    J1: 1/8-inch miniature jack socket
    J2: 1/8-inch miniature jack socket
    S1: single-pole, single-throw miniature switch
    Power supply: nine-volt battery

Project 25: Passive Five-Step Voltage Indicator

Introduction

As we've seen before, there are some test setup cases where an oscilloscope and function generator are essential to verify the correct operation of a circuit under test. But more often than not, a rough indication is all you need, especially in the case of verifying a basic dc voltage level. Even though a simple multimeter is not that expensive, there are times when you might need another (multimeter), to supplement the ones (you have) that are already tied up in a measurement circuit. This passive five-step voltage indicator gives you that capability, by allowing you to get a rapid confirmation of the presence of a dc voltage.

Circuit Description

This five-step voltage indicator is a compact array of five LEDs; it provides a nice, inexpensive means of getting fast voltage information. A string of five resistors (R1, R2, R3, R4, R5) provide the tap off points for five LEDs (D1, D2, D3, D4, D5) respectively. The schematic is shown in Figure 8-5.

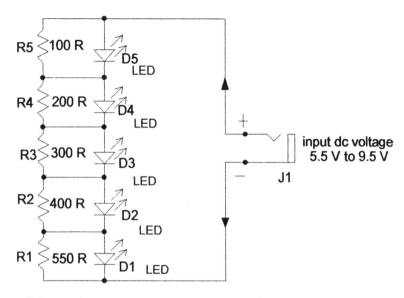

Figure 8-5 Passive Five-Step Indicator

The unknown test voltage is applied across the outer resistor chain. The resistor values are: R1 = 550 ohm, R2 = 400 ohm, R3 = 300 ohm, R4 = 200 ohm, and R5 = 100 ohm. The total resistor chain value is 1550 ohms. This circuit works best for straight dc voltages between 5 to 10 volts or a wide pulse width that gives you sufficient viewing time for the LEDs, with a repetition rate that is sufficiently slow. At about a starting applied voltage of 5.5 V, the first LED1 (D1) will begin to turn on. For higher values of applied voltage, further LEDs in turn (from D1 to D5) will turn on. The approximate input voltage levels at which the LEDs turn on are:

LED1 (D1): 5.5 V
LED2 (D2): 6.5 V
LED3 (D3): 7.5 V
LED4 (D4): 8.5 V
LED5 (D5): 9.5 V

With all the LEDs on, the current drawn is about 20 milliamps, so the circuit is best used for intermittent testing. If you have many nine-volt batteries around, in various stages of freshness, you can easily sort them into various groups of freshness, as it is a simple matter to distinguish between different levels of LED brightness. The resistor values can be varied a little without much consequence, as in practice it doesn't matter what the turn-on voltage values are, so long as the LEDs incrementally turn on in sequence as the input voltage is increased. If you're going to experiment with different resistor values, note that the resistor values should decrease in value when going from R1 to R5.

Construction Tips

A small section of assembly board is all you're going to need for this project. Most of the components are going to need a project case for mounting, though. There are five LEDs to be mounted on the front panel. The input is through a jack socket, J1, which is the usual 1/8-inch miniature type.

Test Setup

A variable dc source (such as a power supply) is ideal for verifying that the individual LEDs turn on as the input is incremented through to about 10 volts. After that you can use a function generator's output square wave (use a low repetition rate) and vary the amplitude to see when each LED turns on. An oscilloscope can doubly verify the actual amplitude at which the LEDs turn on.

Parts List

Resistors (all resistors are 5 percent, 1/4 W)
    R1: 550 ohm
    R2: 400 ohm
    R3: 300 ohm
    R4: 200 ohm
    R5: 100 ohm
Additional materials
    D1: LED
    D2: LED
    D3: LED
    D4: LED
    D5: LED
    J1: 1/8-inch miniature jack socket

Project 26: Active Five-Step Voltage Indicator

Introduction

The previous circuit introduced you to the concept of having a series chain of resistors acting as a potential divider. The effective voltage across each section of the divider chain can thus be monitored and an indication given of the presence and magnitude of the applied voltage. When LEDs are used as monitors and placed in parallel across the resistors, there is going to be a shunting effect by the LEDs. In the LED's off state, its resistance is high, and in its on state, it has a low resistance. Calculating the shunting effect of the LED is thus not straightforward. Fortunately, it is not a requirement. This active five-step voltage indicator circuit, however, is much cleaner to analyze as the use of integrated circuit buffers enables the potential-divider property of the resistor chain to be calculated without any shunting effect. As the applied input dc voltage increases in magnitude, at each one-volt incremental step a corresponding LED will turn on, thus providing us with a nicely controlled voltage monitor.

Circuit Description

The active five-step voltage indicator is a combination circuit consisting of five separate threshold detectors. The schematic is shown in Figure 8-6.

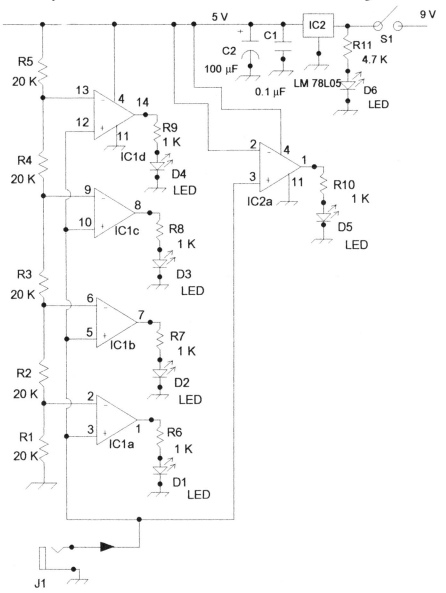

Figure 8-6 Active Five-Step Indicator

A series resistor chain of five resistors generates the five separate reference voltages. Through the use of integrated circuits, each threshold circuit is isolated from each other and hence can be individually configured (not as in the previous circuit).

There are five ICs used in this design, utilizing the LM 324 quad op-amp. Two LM 324s are thus required. IC1a, b, c, d uses the entire second quad set. IC2a (LM 324) uses just one of the four available quads. A potential divider five-resistor chain is formed from five 20 kohm resistors for R1, R2, R3, R4, and R5. The total value is $5 \times 20\,\text{kohm} = 100\,\text{kohm}$. The supply voltage is five volts. Each tap off point produces the following voltage reduction. Across resistor R1, we have (R1/Rtotal) × 5 volts. Thus:

Voltage across R1 = (20/100) × 5 = 1 volt
Voltage across R2 = (40/100) × 5 = 2 volts
Voltage across R3 = (60/100) × 5 = 3 volts
Voltage across R4 = (80/100) × 5 = 4 volts
Voltage across R5 = (100/100) × 5 = 5 volts

These reference voltages are fed to the inverting inputs of each of the five op-amps with the following connections:

Voltage across R1 fed to IC1a/pin #2
Voltage across R2 fed to IC1b/pin #6
Voltage across R3 fed to IC1c/pin #9
Voltage across R4 fed to IC1d/pin #13

The reference voltage for IC2a/pin #2 is taken directly from the five-volt rail. The dc input voltage is fed commonly to the noninverting inputs of:

IC1a/pin #3
IC1b/pin #5
IC1c/pin #10
IC1d/pin #12
IC2a/pin #3

Signal outputs are taken from:

IC1a/pin #1
IC1b/pin #7
IC1c/pin #8
IC1d/pin #14
IC2a/pin #1

Each op-amp operates in the same way. When the signal input exceeds the reference voltage, the output switches from a logic low to logic high state. Five individual LEDs (with associated limiter resistors) placed across the outputs will turn on in turn. These are:

IC1a: LED (D1), resistor R1 (1 kohm)
IC1b: LED (D2), resistor R2 (1 kohm)
IC1c: LED (D3), resistor R3 (1 kohm)
IC1d: LED (D4), resistor R4 (1 kohm)
IC2a: LED (D5), resistor R5 (1 kohm)

Power for this circuit is supplied from a five-volt regulator (IC2). Unlike previous circuits where the five-volt supply is dictated by the specific characteristics of the IC in use, this circuit requires a stable voltage supply (five volts is convenient), since the potential divider requires a stable voltage in order for the thresholds to be constant. If we were to use a nine-volt battery as the source, as the battery aged, the voltage would fall, and hence the reference voltages would change; thus our frame of reference would be lost. A five-volt source is also compatible with the interface from TTL logic and (with the resistor values chosen) it also produces a nice, easy means to calculate one-volt incremental steps in thresholds. Power switch S1 feeds in the nine-volt battery voltage across the supply monitor of LED D6 and resistor R11 (4.7 kohm). The five-volt regulated output is smoothed by capacitors C1 (0.1 µF) and C2 (100 µF). The input terminal is terminated in a 1/8-inch miniature jack socket (J1) to facilitate connection into this circuit.

Construction Tips

Choose a fairly large project case for this design, as there are two big 14-lead DIL IC sockets needed, in addition to the six LED indicators. The case size will be dictated by the size of the assembly board you finally land up with. Start with laying out the resistor potential divider chain of R1, R2, R3, R4, and R5, as the tap-off points will need routing to diverse locations at the IC sockets. The IC sockets can be located side by side once you've optimized the resistor chain. Take care with the layout, as there are many connections to be routed across the assembly board. Apart from the differences in pin connection numbering, each of the five circuits are exactly the same. The greatest chance for error during the construction phase will arise when making the connections to the IC's inverting and noninverting input terminals. These lie adjacent to each other and can easily result in inadvertent placement errors. One way of avoiding the possibility of error is to take note of the "mirror image" arrangement in which the LM 324's op-amps terminals are numbered. Notice that each op-amp is laid out in a symmetrical pattern. This pattern will be a good way of checking your wiring accuracy as each stage is completed. An alternate scheme is to connect up all the inverting input terminations first—these will go to the resistor taps. Then connect up the output pins—these are conveniently located at the four corners of the IC. What's left is the noninverting pins—which are all paralleled up.

Test Setup

The first test will be to verify with a multimeter that you have the correct dc voltages existing across the resistors junctions: R1 and R2 = 1 volt, R2 and R3 = 2 volts, R3 and R4 = 3 volts, and R4 and R5 = 4 volts. Across the entire resistor chain you should, of course, have the starting five-volt condition. Once this has been checked out, any dc voltage feed can be used to test out this circuit, but a slow-moving triangle wave with a TTL format (zero to five volts) is particularly useful, as you can then see the LEDs linearly switch on in time. A function generator source is an ideal test signal source, as it has the capability to provide the necessary test input waveforms. Each of the consecutive LEDs, from D1 to D5, will turn on in turn when the appropriate one-volt increment reference thresholds are exceeded.

Parts List

Semiconductors
    IC1: LM 324 quad Op amp
    IC2: LM 324 quad Op amp
    IC3: LM78L05 5 V regulator
Resistors (all resistors are 5 percent, 1/4 W)
    R1: 20 kohm
    R2: 20 kohm
    R3: 20 kohm
    R4: 20 kohm
    R5: 20 kohm
    R6: 1 kohm
    R7: 1 kohm
    R8: 1 kohm
    R9: 1 kohm
    R10: 1 kohm
    R11: 4.7 kohm
Capacitors
(All nonpolarized capacitors disc ceramic)
(All electrolytic capacitors have a 25 V rating)
    C1: 0.1 µF
    C2: 100 µF
Additional materials
    D1: LED
    D2: LED
    D3: LED
    D4: LED
    D5: LED
    D6: LED
    S1: single-pole, single-throw miniature switch
    J1: 1/8-inch miniature jack socket
    Power supply: nine-volt battery

CHAPTER **9**

Special Power-Supply Circuits: Projects 27–38

Real, practical circuits as featured in this book differ considerably from application notes that are commonly found in semiconductor manufacturer's data books. There are many considerations that separate a real circuit from a suggested application note. This chapter puts together a collection of such factors. Don't let their simplicity deceive you as to the significance of these designs. Usefulness is not the exclusive domain of complexity! Radio frequency (RF) circuits are neither analog nor digital but are often featured in circuit compilations, so for completeness, references are included here too for these designs. Later chapters will feature two RF designs.

Project 27: RF Oscillator Stable 6.2-Volt Supply

Introduction

High-frequency radio frequency (RF) oscillators are often featured in circuits that describe themselves as short-range FM (frequency-modulated) transmitter circuits. One of the key requirements of these circuits, which usually operate in the broadcast FM broadcast band (88 to 108 MHz) is a need for the carrier frequency to be stable. There is little value in having a circuit that drifts in frequency—the end result after your careful build will be less than satisfactory. A little-emphasized fact with this type of circuit is that the stability of the power-supply voltage controls the stability of the oscillator frequency. Stabilize the supply voltage, and the drift in oscillator frequency will be improved. As a battery source will reduce over time as the circuit draws current, the first requirement is to have a line-generated dc power supply. This can be easily derived from a dc adapter that is specified to deliver, say, 15 volts dc. However, most inexpensive voltage adapters of this type are notoriously unsuitable for any serious application. The greatest limitation with these adapters is that the rated output voltage is heavily dependent on the load resistance. The internal circuit in one of these line adapters comprises nothing more than a transformer, rectifier, and a smoothing capacitor, as you could easily confirm if you were to disassemble one of these units. Principally they're meant to power noncritical devices, like calculators. However, we can easily convert this limited output into an extremely valuable function, by the addition of a regulator IC. The basic regulator IC circuit, as we've seen many times earlier, merely takes a varying dc supply and converts it to a fixed and very stable output. For example, a nine-volt battery decreases with usage over time but can be converted into a very stable five-volt supply (for TTL circuits). However, there's another subtle aspect to consider when RF circuit stability is required in designs where a number of circuits, such as an oscillator and power amplifier, are coupled together to share the same power supply. RF oscillators are notorious for not wanting to share—they need their own power supply. But there's no need to have a totally separate power supply. All you need is a separate voltage regulator feeding just the RF oscillator.

Circuit Description

Our circuit here is a cascaded power-supply arrangement, with each successive stage of regulated voltage being less than the previous one. The schematic is shown in Figure 9-1.

We start with an unregulated supply of +15 volts. Typically you'd get this from a so-called "12-volt" adapter. This is one of the popular general-purpose units you can find in any electronics store, used for powering calculators and other noncritical pieces of electronic equipment. When you measure the off-load voltage from the regulator's output, it's actually going to be around 15 volts. The measured voltage is always in excess of the stated voltage. The first regulator, IC1 (LM78L12) in our schematic, produces a nice, clean, stable 12-volt output when fed with the 15-volt as the input. This 12-volt regulated voltage can be used to feed the less critical part of the circuitry that is other than the RF section. The actual voltage needed will depend on the type of

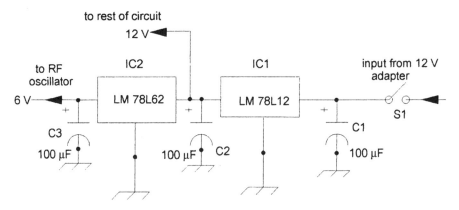

Figure 9-1 RF Oscillator Stable Supply

circuit you're working with. A nine-volt regulator can also be used here, but they're less commonly available than the 12-volt type.

The second regulator, IC2 (LM78L62), has to be a lower-voltage type, shown here as a 6.2-volt type, as its feed voltage is going to be the regulated output voltage from the previous stage. We are essentially going to have two stages of regulation here. This second, twice-regulated supply feeds the oscillator circuit exclusively. With this arrangement, any fluctuations in the rest of the circuitry, fed by the 12-volt regulated source (i.e., from varying load current effects), will not be fed back to the oscillator, and the oscillator's critical frequency is thus protected from drifting. Capacitors C1, C2, and C3 (all 100μF values) are those components usually found in regulator circuits, and they are included to stabilize the input and output voltages. These are located at the input to IC1, the output from IC1/the input to IC2, and the output from IC2.

Both of the voltage regulators (LM7812 and LM7862) come in TO-92 plastic packages. When viewed from the front, with the leads facing toward you and with the flat surface facing upward, the pin identification is:

Left terminal: output
Center terminal: ground
Right terminal: input

For the LM7812, the continuous output current is rated at 100 milliamps. The minimum value of input voltage required to maintain a regulated output is 13.7 volts. Maximum specified input voltage is 35 volts. The regulated output voltage is 12 volts.

For the LM7862, the continuous output current is rated at 100 milliamps. The minimum value of input voltage required to maintain a regulated output is 7.9 volts. Maximum specified input voltage is 35 volts. The regulated output voltage is 6.2 volts.

For ease of connection, three jack sockets provide the feed (J1) and takeoff points (J2, J3).

Construction Tips

Ultimately this circuit is intended to be integrated into an existing design, unlike previous circuits, which are more interface projects. For the purpose of demonstration, however, we can show how this circuit provides the regulating function. The assembly board needs only to be large enough to hold a handful of components—there's very little to it. Label the three jack sockets (J1, J2, J3) to differentiate their functions.

Test Setup

The test setup will demonstrate the range of variation in load voltage this circuit is able to tolerate before the regulating control breaks down.

1. Measure the direct off-load voltage from the 12-volt regulator (taken from jack socket J1). This is going to be in the region of 15 volts. Temporarily shunt the output voltage with a range of load resistors. The absolute values are not important; use whatever values you have to hand. You're looking to see the effect of drawing different load currents on the output voltage. Suggested values of resistors are, 100 kohm, 10 kohm, 1 kohm, and 100 ohm. The output voltage will drop successively lower with lower values of load resistor, showing the unsuitability of the direct use of the line adapter.
2. Next take the voltage reading from the output of IC1 (the 12-volt regulator). This is via jack socket J2. The unloaded output will be a stable 12 volts. Repeat the shunting sequence with the same value of resistors. The voltage will now remain constant or drop negligibly, if at all, in the vicinity of the lower shunt resistor value. The minimum load resistor necessary to maintain a regulated supply is 12 V/100 milliamps = 120 ohm. With a current drawn of 100 mA and a 120-ohm resistor, the power rating of the 120 ohm resistor is 1.2 watts. Temporarily, a 1/4-watt resistor will do before it starts to heat up, so momentarily place it across the output from IC1. A quick indication is all you need of the regulated output voltage.
3. Measure next the output from IC2 (taken from jack socket J3). This will be 6.2 volts. Again place the shunt resistors across the output of IC1, which is also the input to IC2. The output from IC2 remains constant, as changes at the input to IC1 are not translated to the output of IC2. The input to IC2 can drop as low as 7.9 volts. The minimum load resistor to maintain a regulated supply is 6.2 V/100 milliamps = 62 ohm. With a current drawn of 100 mA and a 62-ohm resistor, the power rating of the 62-ohm resistor is 0.62 watts. This is less critical than the previous case, but still, place the shunt resistor long enough to get a reading from the output of IC2.

Parts List

Semiconductors
    IC1: LM78L12 regulator
    IC2: LM78L62 regulator
Capacitors
(All nonpolarized capacitors disc ceramic)
(All electrolytic capacitors have a 25 V rating)
    C1: 100 µF
    C2: 100 µF
    C3: 100 µF
Additional materials
    J1: 1/8-inch miniature jack socket
    J2: 1/8-inch miniature jack socket

J3: 1/8-inch miniature jack socket
12-volt line adapter

Project 28: Stable Five-Volt Source

Introduction

Battery power is great for project portability, but where you've got a circuit that's going to remain on for extended periods (days on end), the battery will eventually run down, especially if the load current is significant (more than several tens of milliamps). A line adapter is a good, rough source of continuous voltage—it'll never run down. But the lack of voltage stability—it's heavily affected by the load resistance it's feeding into—means it's not suitable for any TTL circuit that requires a very tightly maintained supply voltage. This stable TTL five-volt source makes a nice stand-alone power circuit project that is always ready to power up your TTL prototypes. In cases where you've got many TTL circuits cascaded, the total current consumption could be relatively high. Under those conditions the correct circuit operation will be degraded if the supply voltage drops to below five volts, as TTL circuits require a very tightly controlled five-volt supply voltage.

In practice I've found that long-term experimenting with electronic circuits soon drains a battery source, and that'll cause no end of problems. More often than not, I've found, the cause of a circuit malfunction is a weak battery, one that measured initially a solid nine volts but soon drained under load (which invariably occurs during the time I'm not monitoring the circuit supply voltage!). It's happened so frequently that I've taken to label and segregate batteries into good, average, and bad groups, so as not to waste any more misguided troubleshooting. This strategy works. So a good, guaranteed, stable source is essential, whether it's going to be a fresh battery or otherwise. This "otherwise" is what we're describing here. In keeping with the way I structure my books, I always prefer to separate out circuits, even if they appeared previously already as parts of other circuits. There's nothing so frustrating, I find, than stripping out part of a circuit and then wondering if you've left anything out. So if you've already noticed that you've seen this circuit previously in the power-supply section of earlier circuits? You have, but that's the reason why.

The three-terminal, inexpensive LM78XX series voltage regulator is really a remarkable part of the hobbyist's tool kit. In the bare-bones case, no other component is needed to get your regulated stepped-down voltage. Nothing could be simpler. Integrated circuits have simplified the circuit design process significantly, as the equivalent discrete-parts build would be quite an exercise in itself. Where the current requirements are moderate, say, up to 100 milliamps, there's nothing to beat these plastic devices. Most IC circuits rarely draw more than 100 milliamps, so it's very unusual to find these devices being too weak.

Circuit Description

The schematic is shown in Figure 9-2. The feed voltage comes from a dc line adapter rated at supplying nine volts, but that in its unloaded condition is going to be closer to 12 volts, driving a five-volt regulator, IC1 (LM78L05). There's an input capacitor, C1 (100 µF), located after the power switch, S1. IC1 is a three-terminal device, so there'll be a ground connection also. The terminations are as shown. On the output side, there are two further capacitors, C2 (0.1 µF) and C3 (100 µF). The regulated five-volt supply terminates in a 1/8-inch miniature jack socket, J1, to make feeding into an external circuit easier. A monitor LED (D1) and current-limiter resistor, R1 (4.7 kohm), are also situated on the output feed side of the circuit to show when the five-volt supply is on.

The LM78L05 regulator IC comes in a TO-92 plastic package. When viewed from the front, with the leads facing toward you and with the flat surface facing upward, the pin identification is:

Left terminal: output
Center terminal: ground
Right terminal: input

Construction Tips

A project case is needed to house this stand-alone unit, especially as it is intended for usage as an external TTL power supply. The output feed is through a jack socket. If you wish, the input feed from the regulator can be also through a similar jack socket—it makes the unit a little neater looking not to have the line adapter attached to it. Mark up the front panel so that there is no confusion between the input and output.

Test Setup

The test setup requires no more than applying power to the unit, by plugging in the line adapter and measuring the output voltage, which, of course, should be five volts.

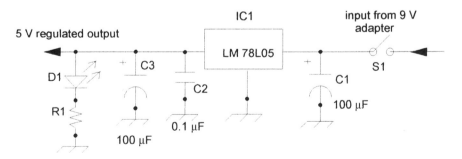

Figure 9-2 Stable Five-Volt Supply

Parts List

Semiconductor
   IC1: LM78L05 regulator
Resistors (all resistors are 5 percent, 1/4 W)
   R1: 4.7 kohm
Capacitors
(All nonpolarized capacitors disc ceramic)
(All electrolytic capacitors have a 25 V rating)
   C1: 0.1 µF
   C2: 100 µF
Additional materials
   J1: 1/8-inch miniature jack socket
   9-volt line adapter

Project 29: Battery or Adapter Supply

Introduction

If you want the portability of battery power and also the sustainability of adapter power, try building this neat arrangement for using either battery or adapter options. To cover a wide range of current applications, the adapter chosen is a nine-volt device with a 300-milliamp capability. This current capacity is more than enough for almost any application you're going to encounter.

Unlike many general-purpose power-supply builds, this one depends critically on getting the mechanical component mix correct (as described in the following section), so be especially careful here.

Circuit Description

The particular device selected for the line adapter is the Radio Shack device (part number 270-1560), with a nine-volt, 300-milliamp specification. There is a special reason for this. Unlike with other adapters, where the specifications may not be available, you have here a device about which there is some data (i.e., from the catalog). You get a selection of output connectors supplied with this adapter. To match the adapter, you need the specific coaxial dc power jack socket, J1 (part number 274-1565). Any one of the many supplied output connectors that comes with the adapter will fit this jack socket, but it is specifically the one that is 2.5 mm in size that you need. To make absolutely sure, identify the one that fits the jack socket absolutely snugly. Be careful of the 2.8 mm size, which appears also to fit at first glance, but because of its larger size it really only makes an intermittent contact. There is also a voltage coaxial polarity selector supplied. Push the coaxial connector into the adapter termination, so that the "neg" marking lines up with the "tip" marking. This is important, otherwise the output polarity will be incorrect! The schematic in Figure 9-3 shows the connections for having a nine-volt battery or adapter option, with a positive Vcc. Take very careful note of the way the terminals are

160 BEGINNING DIGITAL ELECTRONICS THROUGH PROJECTS

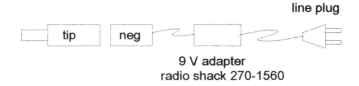

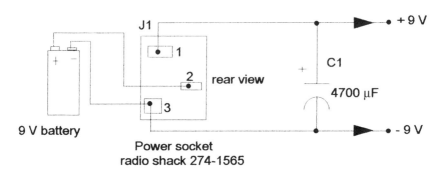

Figure 9-3 Battery or Adapter Supply

identified. With this build there are many unusual mechanical variables that can throw you for a loop. That was the major problem I found myself when putting this build together! The capacitor C1 (4700µF) is a very large electrolytic, selected to remove the significant inherent low-frequency 60 Hz "hum" that would be otherwise transferred to your circuit. For audio applications, this would be annoying. When the adapter connector (female) is not inserted, pins #1 and #2 on the power jack socket are shorted together by the internal makeup of this part. Under those conditions, the positive nine-volt terminal is connected directly to the external load. Pin #3 is the negative or ground terminal. When the adapter connector is pushed into the jack socket, pins #1 and #2 are now open circuit, and pin #1 is now connected to the load; that is, the positive adapter feed now goes to the load. Verify carefully that all polarities are correct before connecting up your load to the power socket. The view shown of the power socket is from the rear.

Construction Tips

The power socket is the key component in this build and requires some customizing to the project case to clear the solder connections. Two small mounting holes are provided for securing the socket to the case. One of the mounting nut/bolt kits available from Radio Shack will fit these holes. In practice what you'll be doing is mounting this configuration in the same case that you're using for your circuit, rather than building this project as a stand-alone unit.

Test Setup

Use a voltmeter to verify the correct polarities are available across the output terminals of the power socket, before any load is connected. After that, there's nothing else to worry about. With the correct operation verified, your circuit will be powered off the internal nine-volt battery when the adapter jack plug is absent. Insert the jack plug, and the power source will be immediately transferred to the adapter.

Parts List

Capacitors
(All nonpolarized capacitors disc ceramic)
(All electrolytic capacitors have a 25 V rating)
    C1: 4700 µF
Additional materials
    J1: power jack socket (Radio Shack, 274-1565)
    9-volt line adapter (Radio Shack, 270-1560)

Project 30: Diode Bridge Circuit Protector

Introduction

Integrated circuits are not forgiving of accidental power-supply reversals. ICs require a positive supply voltage. It is not uncommon to reverse accidentally the supply connections in a moment of haste, and that could be a costly mistake. What can we do? The circuit shown here is a bulletproof protector! With this circuit in tow, there is no way you can transfer an incorrect polarity down to the load.

Circuit Description

The protector circuit is arranged in a four diode bridge configuration as seen in Figure 9-4.

Diodes D1 to D4 are all rectifier diodes (1N4001 type). A diode will conduct when the anode terminal (the back end of the "arrow" symbol) is supplied with a positive voltage. Consider, therefore, the following scenario. With reference to the diode arrangement shown in the schematic, assume first of all that a positive voltage is applied to junction A and a negative voltage is applied to junction B. Diode D1 will conduct because it will be forward biased by the positive voltage applied to its anode terminal. At the same time, the negative voltage applied to junction B will turn on diode D2. Diodes D3 and D4 are reversed biased by this polarity scenario, and it's as if they were not in the circuit—as in the reversed biased position, the diode has a very high resistance. Therefore, across the load terminals, X and Y, we'll find a positive voltage across the X terminal and a negative voltage across the Y terminal.

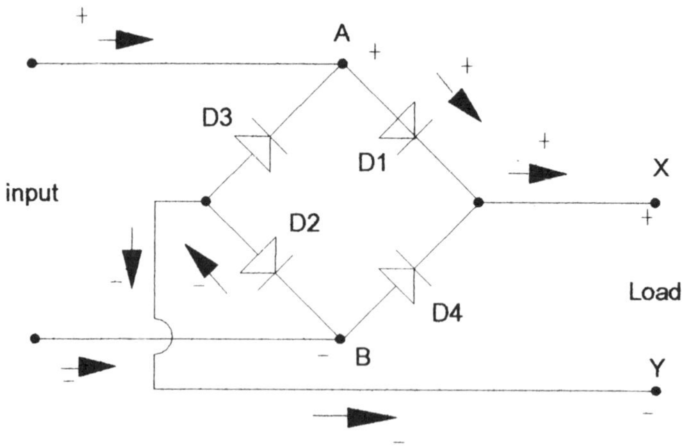

Figure 9-4 Diode Bridge Protector

In the second scenario, we'll assume the situation is reversed: that a negative voltage is applied to junction A and a positive voltage, therefore, applied to junction B. This time around, diode D3 is forward biased, as its cathode terminal has a negative voltage applied to it. Diodes D1 and D2 are reversed biased and do not play a part in the proceedings—they may as well not be there. At the lower end of the bridge, diode D4 conducts also, as it's forward biased by the positive voltage applied to its anode terminal. The result of this is a positive voltage applied to output terminal X and a negative voltage applied to terminal Y. So what we have with this circuit is always a positive voltage available at the output X terminal, regardless of which way the input voltage is polarized—that is, we've effectively safeguarded the output from ever reversing in polarity—irrespective of which way the input voltage is applied.

Construction Tips

The only area to be careful of is to make sure the diodes are correctly connected in the circuit, with regard to their polarity requirements.

Test Setup

Verify before applying the voltage to an actual circuit load, by placing a voltmeter across the output terminals X and Y, that the voltage is as you expect. The voltage should be positive at the X terminal.

Parts List

D1: 1N4001 rectifier diode
D2: 1N4001 rectifier diode

D3: 1N4001 rectifier diode
D4: 1N4001 rectifier diode

Project 31: Regulator Protector

Introduction

The three-terminal regulator, as we've seen from previous circuits, is an extremely simple to use device, providing a wealth of useful circuits. A common circuit arrangement involves having a capacitor and LED monitor across the input and a capacitor across the output terminals. Under certain conditions, the regulator can be damaged if the voltage at the output terminal rises above that at the input. This condition can arise under certain switching conditions where the input capacitor discharges through the input LED indicator. This simple regulator protector safeguards the regulator IC itself.

Circuit Description

The normal circuit configuration for a regulator is given by IC1 (LM78L05), power switch S1, and LED (D1) acting as the power monitor with current limiting resistor, R1 (1 kohm). Smoothing capacitors C1 (10µF) and C2 (10µF) appear across the input and C3 across the output terminals, respectively. The protection comes in the form of a rectifier diode D2, wired across the input and output terminals as shown, with the cathode end going to the input side, as seen in Figure 9-5.

Under the correct, normal operating conditions, the voltage on the input side is greater than the output side. That means that the cathode end of the diode D2 is positive with respect to the anode. The diode is thus reversed biased and is effectively not in circuit because of its high reverse-bias resistance. When the input side drops under abnormal conditions to a level less than the output side, the protection diode's cathode is now negative with respect to its anode. Diode D2 now conducts, thus shunting the

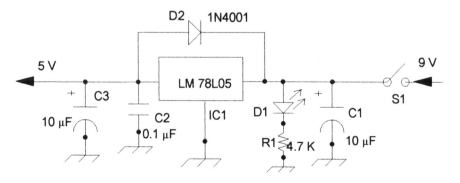

Figure 9-5 Regulator Protector

abnormal voltage away from the regulator IC. This is how the protection circuit works.

Test Setup

There is no testing that can be conveniently carried out. The diode acts as a protector, and its inclusion is all that is needed.

Parts List

Semiconductor
    IC1: LM78L05 regulator
Resistors (all resistors are 5 percent, 1/4 W)
    R1: 1 kohm
Capacitors
(All nonpolarized capacitors disc ceramic)
(All electrolytic capacitors have a 25 V rating)
    C1: 10 µF
    C2: 10 µF
    C3: 10 µF
Additional materials
    D1: LED
    D2: 1N4001 rectifier diode
    S1: SPST switch

Project 32: Reversible Voltage Source

Introduction

Diodes and transistors are commonly tested by means of an ohmmeter to measure the forward and reverse bias characteristics. This simple test is a good indicator of the goodness of such devices. If the basic tests aren't positive, there's no need to use more extensive test methods. This circuit arrangement shown here facilitates the basic measuring process.

Circuit Description

The circuit is shown in Figure 9-6.

A budget-priced analog multimeter, that is, one that typically measures volts, current, and resistance, can serve as a low-voltage source (typically a few volts) that can be used to check the forward and reverse characteristics of diodes and transistors. On the resistance range, when the multimeter is set to measure resistors, the multimeter is actually applying a voltage to the resistor under test and measuring the resultant current that flows. From Ohm's Law, by

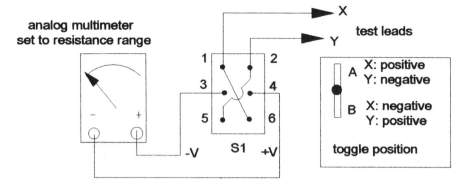

Figure 9-6 Reversible Voltage Source

measuring the resultant current that flows, the unknown resistor value can be calculated. This is the basic principle of measurement for simple ohmmeter circuits. With the specific, low-cost, analog multimeter I'm using, with the meter set to measure resistance on the highest range (in this case the ×1 K scale factor), the measured voltage appearing across the test leads is about three volts, but of the opposite polarity to the color of the test leads. The red test lead is actually the negative voltage source, and the black test lead the positive voltage source—opposite to what you'd expect. Knowing this fact (the "reversed voltage polarities"), you can follow the reasoning behind simple forward/reverse resistance measurements on diodes and transistors. When checking diodes and transistors in order to determine whether a component is faulty or to identify an unmarked device (transistor), you often have to reverse the multimeter leads to what are already small, fragile, and short transistor leads. To avoid a precarious balancing act, wondering if the test leads are in fact making contact, this simple double-pole, double-throw switch is all you need to make cleanly the reversal change to the pins under test.

The switch S1 has three pairs of terminals. The center pair is fed by the source, in this case the voltmeter test leads, with the multimeter set to the resistance range. There is a cross link running across the opposite pair of terminals as shown. The takeoff point for the new external test leads are marked X and Y. The switch toggle positions are marked A (upper) and B (lower). When the toggle faces up (with the switch configured as shown in the schematic), the center terminals make contact with the lower set of terminals. When the toggle is reversed, the center terminals make contact with the upper set of terminals. To eliminate any error, the switch contacts are marked as shown. The connection scheme is:

upper toggle position	pin #4 connected to pin #6	pin #3 connected to pin #5
lower toggle position	pin #4 connected to pin #2	pin #3 connected to pin #1

Construction Tips

A project case is essential, given that there are stable connections to be made between the multimeter to the switch and from the switch to the external test probe leads. Match up the choice of sockets and plugs with what you have on your multimeter. Add a second set for the original tests leads, which will now emerge from the project case.

The connection is as follows. You'll have two plug terminated leads running from the multimeter to the project case, and two test leads emerging from project case. It's a good idea to label the polarities of the voltage appearing across the test leads. Use insulating wire for the cross connections on the back of the switch. If you have a second multimeter, you can use this to verify the polarities of the output voltage.

Test Setup

Couple up multimeter, set to the resistance range, and verify with the two switch positions the polarity of the output voltages.

Parts List

S1: double-pole, double-throw switch
Additional materials
    Basic analog multimeter

Project 33: Jack Socket Tutorial

Introduction

The common 1/8-inch miniature jack plug/socket combinations are found on practically every audio device with a headphone outlet. Which terminals do you use if you're making up your own connections? Sure, you can figure it out, perhaps by trial and error. Personally I like to see the instructions given, so here goes.

Circuit Description

The 1/8-inch miniature jack socket is a component that is most commonly deployed when coupling an audio amplifier to headphones. Because of the fact that commercial audio systems operate in a stereo mode, these jack plugs and jack sockets are stereo components. For simplicity however, the mono versions are used in the description here. The same basic information applies; you just have two of each item. The circuit is shown in Figure 9-7.

For the beginner, it is quite likely that some confusion will arise when buying or using these jack plugs and sockets. First of all, there are two different types of jack sockets available, called "normally closed" and "normally open." When the jack socket is viewed sideways on, you'll see solder connections, one

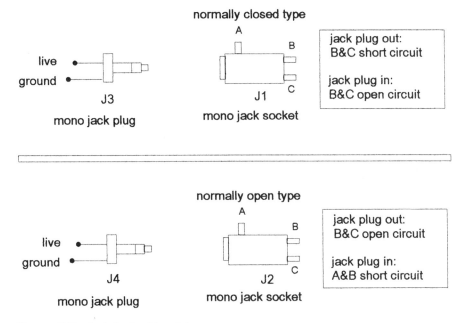

Figure 9-7 Jack Socket Tutorial

facing "north" and two facing "east." Okay, that sets up the identification of the terminals. These are labeled as "A," for the north one, and, going in a clockwise direction, "B" and "C" for the two facing east. The upper A terminal is always connected to ground, and the lowest C terminal goes to the live signal. In the case of an audio power amplifier being coupled out, terminal C would go to the final output capacitor leading off from the amplifier's output. This connection scheme applies to both types of jack sockets.

The difference comes in as follows. With the "normally closed" type of jack socket, terminals B and C are shorted together (i.e., they're "closed") when the jack plug is pulled out (this is the "normal" position). The "normally closed" descriptor refers to the fact that terminals B and C are shorted together when there is no jack plug inserted. When the jack plug is inserted, the B and C terminals are open circuit.

With the "normally open" type of jack socket, terminals A and B are open circuit (i.e., they're "open") when the jack plug is pulled out (this is the "normal" position). The "normally open" descriptor refers to the fact that terminals B and C are open circuit when there is no jack plug inserted. When the jack plug is inserted, the A and B terminals are short-circuited. So, which type of jack socket should you use? If the requirement is for a regular ground and live signal connection, either type will do—the ground connection goes to terminal A, and the live signal goes to terminal C. In both cases, ignore terminal B.

For the special case, however, of a radio having both an internal speaker and a headphone connection, we know that when the headphone jack plug is inserted, the speaker is automatically disconnected. How is this done? For that application, the "normally closed" type of jack socket is used. Terminal A is the ground connection, as we would expect. Terminal C goes to the amplifier output, as once more would be expected. But the difference occurs with the connection to terminal C. One speaker terminal goes to the jack socket terminal C (the other speaker is grounded, as normal). As we've indicated earlier, with the "normally closed" type of socket, terminals B and C are shorted together when the jack plug is not present. This being the case, the radio's internal speaker is activated with the headphone left unplugged. When the headphone jack is inserted, the terminals B and C go open circuit. That means the internal speaker is now disconnected. The radio signal is fed in through terminal C and gets routed through the jack plug out to the headphone. The normal headphone ground gets coupled to the ground terminal A, which remains unchanged regardless of whether the jack plug is absent or present.

The jack-plug half of this set is easier to follow. When the barrel of the jack plug is unscrewed you'll see two solder terminals, a short one and a long one. The shorter terminal is the one that gets connected to the farthest end of the jack-plug tip. This is the connection that gets made always to the live signal. The longer terminal and the more solid-looking of the two terminals is the ground connection. It is electrically connected to the jack plug tip that is closer to the body section. Between the tip portions you'll see a narrow section of insulating material present.

Construction Tips

The leads to the jack socket are fairly fragile, so take care that the flexing in the leads is limited after they've been soldered in. The solder terminals to the jack plug's live terminal is also fragile and should be soldered in only after the more sturdy ground connection is made. That way any undue flexing in the wire is curtailed by the already-made ground connection.

Test Setup

Nothing beats actually seeing the connection scheme yourself. For the test you'll need a jack plug and socket set, and also a multimeter set to the resistance range for checking continuity. With just the bare jack socket, you can verify the continuity between the appropriate terminals, as defined above. Next insert the jack plug and verify that the correct terminals are now connected.

This is the test sequence and the expected results.

Jack plug and "normally closed" jack socket. Continuity test.
1. Jack plug out. Jack socket terminals B and C shorted.
2. Jack plug in. Jack socket terminals B and C open circuit.

Jack plug and "normally open" jack socket. Continuity test.
3. Jack plug out. Jack socket terminals A and B open circuit.
4. Jack plug in. Jack socket terminals A and B shorted.

Parts List

J1: 1/8-inch miniature mono jack socket "normally closed" type
J2: 1/8-inch miniature mono jack socket "normally open" type
J3: 1/8-inch miniature mono jack plug
J4: 1/8-inch miniature mono jack plug

Project 34: Meter Overload Protector

Introduction

If you know and can be sure of the dc voltage being measured, a monitoring analog meter's range can be safely selected to prevent any excess voltage from damaging the meter's movement. Sometimes, however, the voltage under test is exceeded or a connection error suddenly applies a much higher than anticipated voltage. It can happen quite easily. Should it happen, the meter's pointer needle is likely to suffer permanent damage—potentially quite an expensive experience, one that is best avoided if you can. DC analog meters are well suited to verifying dc voltage levels without the need for any further components. Contrast trying to do that with a digital panel meter—there are many components needed to get the same display. An analog meter also doesn't have to be powered up (like digital voltmeters), and dynamic changes can be easily interpreted. However, there is a downside; too high an input voltage, usually inadvertent, will likely send the pointer to a calamitous journey over the end stop. The robustness of a digital meter makes it practically indestructible from voltage overloads, but an easily implemented overload protector circuit will take away the risk of (analog) meter damage.

Circuit Description

The dc input voltage to be monitored is fed to resistor R1 (1 kohm), potentiometer VR1 (10 kohm), and the meter itself, as seen in Figure 9-8.

The arrangement is just a basic potential-divider circuit. VR1 is adjusted to match the maximum input voltage encountered, to the full scale deflection on the analog meter scale. A protection diode is placed in parallel across the lower half of the divider circuit, as shown. For low-magnitude dc input voltages, the signal is fed via R1 and VR1 to the meter, with the diode effectively being bypassed. If a high-overload condition occurs, the diode D1 (1N914) will conduct, thus protecting the meter. The breakdown voltage across D1 is about 0.6 volts. For a higher breakdown value, you can cascade several diodes in series.

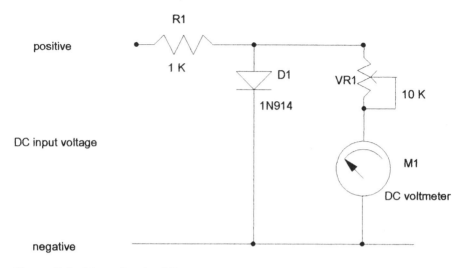

Figure 9-8 Meter Overload Protector

Construction Tips

The meter itself will most likely be a multimeter set to read dc volts. If you are going to use this option, the size of the project case can be quite small, as the case needs house only the potentiometer (the largest component) and two other very small components (R1 and D1). There will also be two sockets required, one for the input voltage and one for feeding to the multimeter itself. Alternately, you can use a stand-alone dc voltmeter itself, in which case the project case needs to fit the meter size. These meters require a very large cut-out to accommodate the rear of the meter. Even with a plastic project case, this is quite a task, and furthermore, the case is going to be quite large to fit the meter. Meters of this type are quite expensive also. My preference is to use a low-cost analog multimeter, which is far more advantageous in all respects of size and cost. Individual analog meters, such as dc meters, tend to be expensive compared to a low-cost analog multimeter; in addition, you get all the additional functions from a multimeter. Connections to a multimeter are via separate banana type plugs. These would match up against the supplied test leads. So, if you purchase a set of connector plugs to match those on the multimeter, these can be mounted on the project case and connected directly to the multimeter. There are several options; either bring out a pair of leads directly from one end of VR1 and ground, or, to make the appearance of the build neater, make up a cable with two matching multimeter plugs on the end. Going with this arrangement you could use a 1/8-inch jack socket on the project—it's neater because of its smaller size. That's my personal preference.

Test Setup

With the build completed, determine what your maximum input dc voltage is going to be and adjust VR1 accordingly so it corresponds to a full-scale deflection on the meter scale. The key factor to note is that the voltage across the protection diode is limited to around 0.6 volts, the breakdown voltage for a silicon diode. This is the voltage above which the diode starts to conduct heavily in the forward bias direction; it is defined as the bias polarity that applies a positive voltage to the anode terminal of the diode. The polarity of the input dc voltage needs, therefore, to be positive with respect to ground. When a dc voltage in excess of the diode breakdown voltage is applied, the diode conducts, and the hazardous voltage is safely shunted away from the meter's fragile movement.

Parts List

Resistors (all resistors are 5 percent, 1/4 W)
    R1: 1 kohm
Additional materials
    VR1: 10 kohm potentiometer
    D1: 1N914 silicon diode
    M1: multimeter set to dc volts range

Project 35: Jack Socket-Power Switch

Introduction

There's nothing as satisfying as making a component do double duty and getting something for nothing! This circuit idea features getting a standard jack socket to act as both a feed source into an amplifier and also to switch on the power, at the same time. By pulling out the jack plug you not only disconnect the signal source but shut off the power supply—clever!

Circuit Description

This circuit as shown in Figure 9-9 will only work with the "normally open" type of jack socket. As we've seen earlier, the interconnection scheme for this particular socket is given by:

1. Jack plug out: Jack socket terminals A and B open circuit.
2. Jack plug in: Jack socket terminals A and B shorted.

where A is the ground connection, B is the auxiliary terminal, and C is the live terminal.

The key feature we're making use of here is that two jack socket terminals, A and B, are shorted only when the jack plug is inserted; normally these terminals would be open circuit. We can make use of this feature by placing terminals A and B in series with the ground connection in a power-supply circuit. The connection from the battery's negative (ground) terminal goes to the jack socket's terminal A, with terminal B going to the circuit's ground. The jack socket now essentially acts as a switch—except it's in the ground line. Let's say you have an

172 BEGINNING DIGITAL ELECTRONICS THROUGH PROJECTS

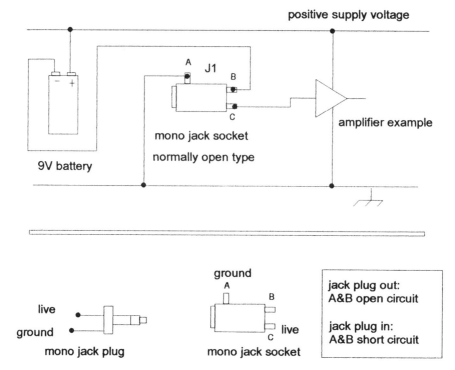

Figure 9-9 Jack Socket-Power Switch

amplifier circuit as shown in the example schematic; you'll see that we have an arrangement where the jack socket is taking the place of the usual power switch. It's actually switching in the ground connection as opposed to the customary switching in the live (positive) line, but that doesn't make any difference.

Without a jack plug inserted, the amplifier circuit has no power applied. With the jack plug inserted, the ground connection for the power supply is completed, and the amplifier turns on. That's the power-switching part of the circuit taken care of. At the same time the signal feed (from the jack plug) is applied to the jack socket's live terminal, C. In one action of inserting the jack plug, therefore, you have both the power turned on and the signal fed into the amplifier input. We save having to have a separate power-supply switch; where space is a premium, this could be a useful advantage.

With the larger 1/4-inch type of jack socket/plug you can use the same trick for feeding an electric guitar signal into a guitar amplifier. These 1/4-inch components are larger, more robust, and easier to see and work with.

Construction Tips

Ideally, you'll need a circuit such as an amplifier to verify the design here. Include an LED monitor in the power circuit so you'll know when the power is applied. The input jack socket, J1, is mounted on the project case.

Test Setup

Without the jack plug inserted into the jack socket, check that the monitor LED is off, as the ground connection from the battery is incomplete. With the jack plug inserted (regardless of whether a signal is applied or not), the ground connection from the battery will be completed, and the monitor LED turns on. If you have a signal input coupled in through the jack plug, this'll be coupled in to the amplifier's input.

Parts List

J1: 1/8-inch miniature mono jack socket normally open type

Project 36: Zener Five-Volt Supply

Introduction

TTL logic circuits require a tightly controlled five-volt supply, which is most easily supplied from a IC regulator. But if you're stuck for a regulator IC, and you desperately need a five-volt supply, here is a quick alternative that might be more easily configured from parts in your component store. It works best for a simple TTL circuit that draws minimal current.

Circuit Description

The common nine-volt battery useful for powering analog ICs is too much for TTL digital circuits that require a five-volt supply. A two-resistor potential divider might first spring to mind as a swift and easy way to divide down the nine volts to a five-volt level. However, placement of any significant current drawing load will affect (reduce) the unloaded calculated voltage, so that's not satisfactory. A better arrangement is to use a zener diode. The nine volts can be easily stepped down by using just two components, the aforementioned zener, and a current-limiter resistor. Zener diode D1 is a 5.1-volt-rated device that performs this function admirably. A zener diode is a special type of diode that when reverse biased breaks down at the voltage specified for the device (each device is rated at a specific breakdown value)—that is, it conducts heavily. In the reverse bias mode, it is able to sustain quite a high current at a very constant value of reverse-breakdown voltage. This breakdown voltage is what we use to provide a suitable voltage supply for a TTL load. The low-value resistor R1 (100 ohm) limits the current flowing through D1 yet provides plenty of drive current for the TTL load. LED D2 is the conventional supply monitor, with resistor R2 (4.7 kohm) limiting current through D2.

Construction Tips

As there are only two components in this design, they can be easily incorporated into your existing circuit design. All you need to watch out for is the correct polarity of the zener diode. The circuit is shown in Figure 9-10.

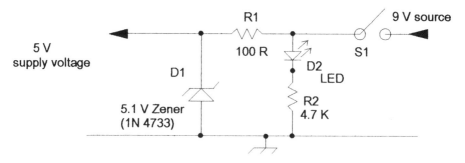

Figure 9-10 Zener Five-Volt Supply

Test Setup

With the setup completed, measure the off-load zener voltage across D1 to confirm that the voltage is 5.1 volts. Once confirmed, connect up the TTL load and remeasure the voltage again to confirm that the voltage is stable at 5.1 volts.

Parts List

Resistors (all resistors are 5 percent, 1/4 W)
  R1: 100 ohm
  R2: 4.7 kohm
Additional materials
  D1: 5.1-volt zener diode (1N4733)
  D2: LED

Project 37: RC Integrator Differentiator

Introduction

Square-wave pulses are quite easily generated, as we've seen from earlier circuits, using typically the LM 555 as a convenient source in the free-running, astable mode. Some applications, especially audio circuits, are more likely to use a sine-wave signal for circuit testing. Sine waves are not as readily generated as square waves—there is no readily available part equivalent to the LM 555. What's the solution? Fortunately, a triangle wave is sufficiently similar to a sine wave, especially when we're looking for a test signal. We can easily convert the stock square wave into a triangle wave shape, using an RC integrator circuit. The complement of the RC integrator is the RC differentiator, which produces a series of pulse spikes, coincident with transitions the originating square wave makes. The pulse spikes can be used as convenient timing indicators in logic circuits.

Circuit Description

The starting point for either of the circuits is a square wave; this is a waveform that rapidly rises in a positive direction, holds that value for a certain length of time, and then drops rapidly to zero, holding that value in turn. The return to the positive direction marks the start of a new cycle. A pulse train such as this that repeats indefinitely is defined as a "periodic waveform." Two commonly encountered techniques for translating the basic square wave are the integrator and differentiator. The terms derive from calculus, where to "integrate" means to find the "area under the function" and to "differentiate" means to determine the "rate of change of the math function."

Since the electronic RC integrator and differentiator circuit's component take on specific values that are dependent upon the characteristics of the input waveform, we'll have to start by defining a specific input waveform. Let's assume we're beginning with a stock 10 kHz audio frequency waveform located comfortably in the audio spectrum. The duty cycle is assumed also somewhere near 50 percent. For both of the circuits described here, the input square wave must be dc coupled.

RC Integrator

The integrator circuit consists of a series resistor R1 (1 kohm) and a shunt capacitor C1 (0.1 µF). The input square wave is fed into resistor, R1, and the output is taken across the capacitor C1. The arrangement resembles a simple potential-divider circuit. These component values are chosen specifically to match the 10 kHz input reference signal. The integrator time constant (RC product) has to be greater than the pulse width for the correct integrating relationship to be maintained. The RC product is defined as the product of R × C. Thus: $R1 \times C1 = 1.10^3 \times 0.1.10^{-6} = 0.1$ milliseconds. If we assume our reference square wave has a 50 percent duty-cycle format, with a 10 kHz frequency, this would translate into an "on" period of 0.05 milliseconds.

The output waveform that is generated across capacitor C1 resembles a triangle wave. Because we have a resistor (i.e., R1) coupling the input signal, the dc component (of the input signal) that we started with is going to be retained in the output (triangle) signal. The triangle wave has several uses, especially as a less harsh version of the original square wave, for use as a quasi-sine-wave test signal generator, or as a ramp generator (especially where the input signal is dropped down to a few Hz), to monitor threshold changes in comparator circuits. When deciding on the value for the integrator components, start with the value for R1. It must not be too high, otherwise the input signal becomes too heavily attenuated. The value for C1 can be fixed afterward.

RC Differentiator

The differentiator has the placement of the circuit components reversed. The signal feed this time goes to a capacitor, C2 (0.001 µF) that is followed by a shunt resistor, R2 (1 kohm). The output is taken across the resistor R2. This waveform is totally different from the previous (integrator) case. At each point in time where the input square wave changes state (i.e., makes a logic transition change), a pulse spike is produced at the output. When the input transition is positive going (from zero to a positive value), there is a positive-going output spike produced. Alternately, when the input transition is negative-going (from a positive value to a zero value), there is a negative-going spike produced. Note that because of the fact the input signal is being coupled through a capacitor (i.e., C2), the dc component we started with is going to be removed—that is, the output waveform sits on the zero-volts rail, and the pulse spikes are either positive or negative in polarity (as opposed to being positive or negative going).

The spike differentiator output is typically used as sharp trigger pulses for locking onto either the positive or negative-going edges of a square wave, or as timing pulses. Since the pulses are placed symmetrically about the zero voltage line, you can use a rectifier diode to select either the positive or negative pulse spikes. The deciding value for resistor R2 is governed by the fact that too low a value will shunt the signal excessively to ground; we choose a value that is not too low (1 kohm is appropriate).

Construction Tips

As there are only two components needed, a resistor/capacitor (for either the integrator or differentiator circuit), there are no special construction precautions to be followed. The circuits are shown in Figure 9-11.

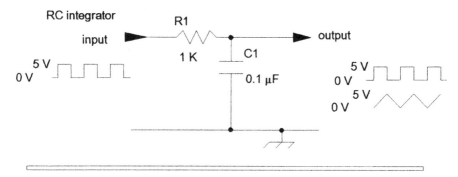

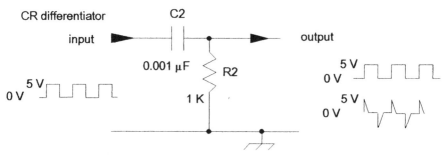

Figure 9-11 RC Integrator Differentiator

Test Setup

To view the effects of the integrator and differentiator, an oscilloscope is required. Couple the input square wave into channel 1 and the modified waveform into channel 2. Take note of the coincidence of the transition points for the translated waveforms, especially where they coincide with changing edges on the input waveform. You can experiment with different RC values to see the effect on the output waveform. The input source is best derived from a function generator, where you've got control over the frequency.

Parts List

Resistors (all resistors are 5 percent, 1/4 W)
  R1: 1 kohm
  R2: 1 kohm
Capacitors
(All nonpolarized capacitors disc ceramic)
(All electrolytic capacitors have a 25 V rating)
  C1: 0.1 µF
  C2: 0.001 µF

Project 38: Variable Resistor Substitute

Introduction

One of the most useful accessory circuits that you can have in your electronics tool kit is a variable resistor substitution source. When experimenting with circuit builds, I've found the need for a range of resistor values to play with. This extremely useful (and yet simple) project has been the most frequently used part of my electronics tool kit to date. It's basically composed of three potentiometers, configured with two flying leads per potentiometer, which can then be temporarily hooked into a circuit under development to determine the optimum final resistor value needed.

Circuit Description

The circuit as shown in Figure 9-12 consists of three potentiometers, VR1 (1 kohm), VR2 (10 kohm), and VR3 (100 kohm), which are wired up with a lead taken from the wiper terminal and one of the outer terminals. When each potentiometer is rotated clockwise, the resistance across the flying leads increases. To identify the correct terminals to use, when you view the front of the potentiometer (i.e., with the potentiometer shaft facing you), the outer terminal is the one located to the right-hand side of the wiper terminal. Each potentiometer is wired up similarly, with two flying test leads emerging.

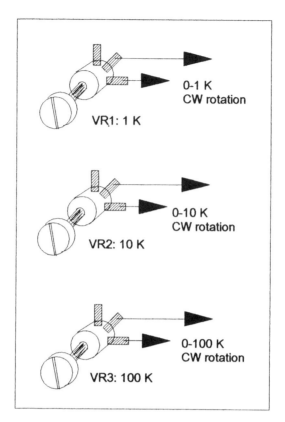

Figure 9-12 Variable Resistor Substitute

Construction Tips

The test leads will be subjected to numerous insertions into the circuit under development, so the first requirement for the choice of wire type is that the emerging leads must be flexible. These flying leads need only be small in diameter—the type of wire used in ribbon cable is perfect. Just strip off two leads per potentiometer—start with a length of 12 inches. This type of stranded wire, though, is very fragile and unlikely to stand up to continual wear and tear, so a better idea is to solder a short length of solid hook-up wire to the end of the flexible flying leads. You need to add a mechanical support to the junction of the solid to the flexible lead, to stop the stranded wire from tearing away when it's in use. Tape down something solid, like a matchstick, across the junction to effect the support bond. Use a project case to mount the three potentiometers and allow sufficient space for a calibration label behind each control knob, for marking up later.

Test Setup

Connect an ohmmeter across each set of potentiometer flying leads and calibrate each of the variable resistor potentiometers in turn, with a range of markings between zero ohms and the full-scale value. Add enough calibration points so that you've got sufficient information on the value of resistance generated across the flying leads.

Parts List

VR1: 1 kohm potentiometer
VR2: 10 kohm potentiometer
VR3: 100 kohm potentiometer

CHAPTER **10**

Single-Chip IC FM Receiver: Project 39

Here's a bulletproof, easy-to-build yet high-quality FM receiver for the hobbyist.

As an electronics enthusiast you've no doubt, like myself, looked at hundreds of receiver circuits in the past, all fired up with enthusiasm and soldering iron at the ready—but to no avail, because the circuit specifies impossible-to-find components, such as special coils and transformers. I started building crystal sets, as I'm sure almost everyone else did, as an introduction into wireless electronics. Crystal sets were extremely rewarding to work with! With so few components, success was almost guaranteed—add a long wire antenna, a ground, and you were in business.

Best of all, in a crystal set there is no battery required—it's the ultimate "getting something for nothing" circuit. Of course, it's limited to the reception of just the AM broadcast band; that's a minor point, but sooner or later you're going to want to advance onto a more challenging receiver. The basic AM detector circuit (which is what the crystal set is) can be updated by adding a power amplifier to drive a speaker load, but that's where the enhancement stops. We haven't actually improved anything in the front end, what is called the radio-frequency (RF) section.

A more advanced type of receiver design can be effected by having some RF amplification take place before the incoming radio signal is detected. This type of technique is employed in the tuned radio-frequency (TRF) type of receiver design. The TRF is not used in consumer designs (i.e., in receivers for domestic use), although it can be a nice type of circuit to experiment with on your workbench. It is basically the radio receiving technique that was the precursor to the next stage of receiver advancement, a more sophisticated design based on what is called the superheterodyne principle, employed by all consumer receivers.

Tuned Radio-Frequency (TRF) Receiver

The TRF receiver has a radio-frequency amplifier located at the front end of the circuit. With the use of a variable capacitor in a tuned circuit, the incoming signal can be selected by resonating the tuned circuit to the frequency you want to receive. After being RF amplified, the signal goes through a detector (or demodulator stage), as is common in any receiver type, to recover the audio information. That audio signal is then fed into a conventional audio amplifier (more often than not using the ever-popular LM 386 audio power IC). The key issue with this type of receiver design is that the first RF amplifier stage must be operating at the frequency of interest. For the AM (amplitude modulation) band, the radio signals transmitted from the broadcast station spans the 550 kHz to 1600 kHz frequency range, which is not too high a requirement when it comes to designing such an RF amplifier. But this would still be a higher requirement than for an audio amplifier that would span typically a 15 Hz to 20 kHz range. For an FM receiver this radio signal span is much higher, lying between 88 MHz to 108 MHz, which poses an even more stringent requirement for the RF amplifier design. With any amplifier design, the higher the frequency demands are, the more difficult and the more stringent the design requirements become. Performance limitations of the TRF receiver essentially limit its usefulness as the basis for commercial receivers. The upgraded alternative is the "superhet" receiver.

Superheterodyne Receiver

The superheterodyne design is the classical receiver technique employed for practically every consumer receiver design. The superhet operating principle is as follows. The incoming radio signal is first amplified by a wide band RF amplifier, and the amplified signal is fed to a mixer stage. The mixer stage, as the name implies, mixes two signals together. There are several signal combinations produced from the mixer output, but (by means of selective filtering) only the difference frequency between the two signal frequencies is used. To get this difference frequency, the mixer stage is fed with a local oscillator (LO) circuit that is ganged to the RF amplifier, so that they always track together. In that way there is always a fixed frequency difference between the two signals (the RF and LO signals). The difference frequency, called the IF (intermediate frequency), is fed to an IF amplifier. For the AM band, the IF is typically 470 kHz, and for the FM it is 10.7 MHz. Having the IF amplifier fixed in frequency makes the design requirements much more controllable, as only a single frequency needs to be amplified. That is the key advantage of the superheterodyne concept. Following the generation of the IF signal, there is a conventional detector stage to remove the high-frequency carrier frequency to leave just the required lower-frequency audio component (where the information is). As this is an easy-to-amplify signal (because of the lower frequency), the audio amplifier requirements are uncomplicated.

Ideally, a superhet receiver would be the preferred design if we were to build a receiver. However, special coil requirements in almost all superhet receiver designs prevent the hobbyist from ever building these designs. Other obstacles include complex alignment procedures requiring expensive equipment and in-depth knowledge. You can, of course, buy a prebuilt mixer or combinations thereof, but this is not the same as building a circuit from scratch. It would seem therefore that other than the ubiquitous crystal set, which is plagued with sensitivity and selectivity limitations, there is very little option except to move onto a less irritating project.

So it was with some trepidation that I tackled this project, knowing beforehand the difficulty of designing a successful receiver build. I had come across a single IC FM radio receiver design, housed in an 18-lead DIL package, that seemed to be quite an interesting device. Would it work? More importantly, what were the coil and alignment requirements? The suggested circuits I had seen specified only one coil, and it seemed to be not too critical. However, it was going to be a try-and-see situation, with the need to build the complete receiver first. This was an FM design spanning the usual broadcast band from 88 MHz to 108 MHz. So I decided to undertake this build. Needless to say, the first designs took a while to get going, as the coil design was basically a trial-and-error affair. But it did work remarkably quickly afterward, and exceedingly well too. That's the reason for the project you see here in this chapter.

The FM receiver IC itself is the TDA 7000. Having built a remarkably good FM receiver out of this device, I can testify that it is a remarkable IC! The receiver is stable, sensitive, and selective. There are no special coils required, other than one noncritical, easy-to-wind home-brew affair. There are also no setup procedures needed—another benefit. Contrary to all the sage advice given on building a high-frequency RF design—keep wiring short, use ground planes, etc.—I decided to evaluate the TDA 7000 deliberately in the worst possible situation, by using a less than tidy prototype board for the assembly platform and making no particular effort to minimize wire lengths. The IC performed beautifully in this mode, and it was impossible to get it to go unstable, testifying to the excellent design of the IC.

This was a very satisfying circuit to design and build. There are also relatively few components needed—mainly a bunch of capacitors. Apart from having to optimize the coil design, the circuit itself worked as soon as it was switched on.

Circuit Description

The TDA 7000 (IC1) is a complete FM receiver in an 18-pin DIL IC (manufactured by Signetics). The total integration of a complete FM receiver in an easy-to-use, single IC has been prevented to date by the need for LC (inductor/capacitor) tuned circuits in the RF, IF, LO, and demodulator stages. Signetics has consequently reduced the normal IF frequency of 10.7 MHz to a much lower frequency of only 70 kHz, which means that active RC (resistor/capacitor) filters can be used as tuning devices. The resultant operational amplifiers and resistors can, of course, be very easily integrated in an IC format. The required capacitors are mounted off board, hence the relatively large number of them found on the schematic. The recommended supply voltage is between 2.7 to 10 volts, so a stable 5-volt supply has been chosen here. A very low supply current of typically eight milliamps would be drawn under those operating conditions. The circuit is shown in Figure 10-1.

Frequency tuning is done with a variable capacitor. As you'll see later, this is done with a very-high-frequency (VHF) varactor diode rather than the usual mechanical trimmer capacitor. A mechanical tuning capacitor is much more cumbersome. It's also larger in size and difficult to obtain. The "varicap" diode tuning arrangement (as recommended here), on the other hand, is very easy to set up and use, and it performs admirably. The actual tuning for the varicap diode is done with a potentiometer that supplies the tuning voltage for the varicap diode. A varicap diode is a special semiconductor device that generates a variable-capacitance output in response to a varying reverse bias voltage. The inductor portion of the tuned circuit is provided for by a simple home-brew coil design. For the antenna all you need is a short 12-inch length of hookup wire. The audio output feed from the TDA 7000 is more than adequate for feeding into a simple LM 386 audio power amplifier (another cool IC for getting a simple and easy medium for driving a low-impedance speaker load).

The input FM signal (88 MHz to 108 MHz) is captured by the short 12-inch antenna, which is capacitively coupled via capacitor C6 (200 pF) to pin #13. The component numbering scheme essentially flows from left to right on the schematic, hence the reason for starting with C6. The frequency tuning components are L1 (defined later in detail), C16 (typically 10 to 20 pF, but the actual value is not critical), D1 (Motorola varicap tuning diode, MV209), C17 (0.01 µF), R2 (100 kohm), C18 (0.1 µF), and VR1 (100 kohm potentiometer). All are coupled as a group into pin #6. D1 is the varicap tuning diode, and VR1 is the potentiometer supplying the dc tuning voltage. The capacitance of D1 varies with the reverse-biased voltage applied to it, and together with the inductor L1 it forms a tuned circuit that selects the required FM frequency. The audio signal is capacitively coupled out from pin #2 via capacitor C11 (0.1 µF). Any audio frequency power amplifier, such as the popular LM 386 IC, can be used to drive a small 8-ohm speaker. The rest of the components shown in the schematic are needed just to support the internal operation of the TDA 7000. These have limited user value, and once in place they can be forgotten as they can be considered as "glue."

Construction Tips

Begin by winding coil L1. This is done by simply taking a length of insulated, solid hookup wire and winding about seven turns on a regular pencil

Single-Chip IC FM Receiver: Project 39

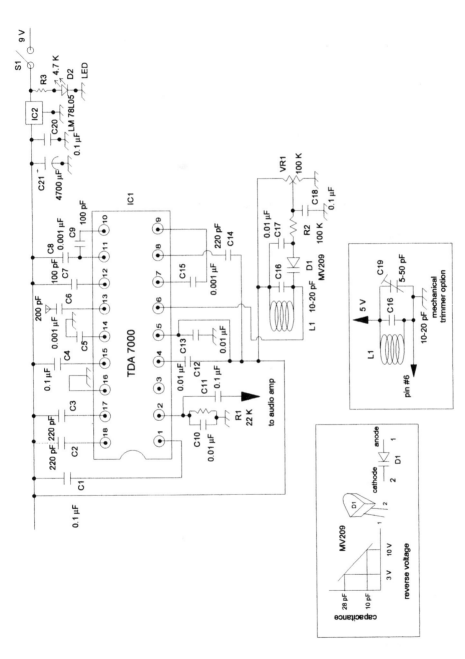

Figure 10-1 TDA 7000 FM Receiver

(used as a former). Remove the coil and space out the turns slightly. My experiments have shown that the coil dimensions are noncritical. I've tried turns ranging from three to nine, formers smaller and larger than a pencil diameter, and with both tight and very loose coil turns. All variants without exception have worked perfectly.

A small solder prototype assembly board is used to attach all the components. Start with the placement, somewhere in the center of the board, of an 18-lead DIL socket for the TDA 7000 IC. For proper orientation, the cutout in the IC socket faces to the left. The IC, when placed in the socket, will have pin #1 located in the far lower left-hand corner of the IC socket. Start with the assembly of the capacitors, and be very careful of the markings on the capacitors, as there are many of them and the physical dimensions and characteristics can be very similar. Ensure that the components are located into the correct holes before soldering, to eliminate difficult removals later if you make a mistake—the capacitors are placed very close to each other, and it is easy to make a mistake. None of the capacitors around the TDA 7000 are polarity sensitive (there are no electrolytic capacitors needed), so that's one item less to worry about. Leave the placement of the tuning components (clustered around pin #6) until later.

The potentiometer, VR1, is mounted off board on the front panel of a plastic case, as it will be subjected to much mechanical usage. It can be calibrated later with markings to identify the FM stations picked up, to facilitate tuning.

The specified varicap tuning diode, D1 (MV209), together with the inductor, L1, will cover the VHF tuning range of 88 MHz to 108 MHz when fed with the appropriate dc tuning voltage (provided by VR1). To identify the correct orientation for D1, hold the diode with the flat surface facing you and the pins facing downward. The left-hand pin will then be the anode terminal, and the right-hand pin the cathode terminal. If you have difficulty in finding the varicap diode, the inset diagram shows the optional connection scheme for a mechanical variable capacitor (C19). The value for C19 is not critical; it can be between 5 pF and 50 pF. The point-to-point wiring is not critical, and again I deliberately stayed away from trying to keep the wires as short as possible, to test the stability of the IC. I found the TDA 7000 an impressive IC to use, with very stable performance. In spite of the deliberate use of less than optimized layout, the circuit performed admirably.

The power-supply requirements of five volts comes from a regulator IC2 (LM78L05) with power monitor LED D2 and current limiter resistor R3 (4.7 kohm). The usual capacitors, C20 (0.1 µF) and C21 (4700 µF), stabilize the voltage. As there's an audio emphasis in this project, it's essential that the hum level be kept to as low as tolerable; that's why the value for capacitor C21 is so large. The feed supply voltage comes from a nine-volt-rated line adapter, as the intent was that the receiver would be operating continually; a nine-volt battery source would have to be replaced too frequently. The power connection is to pin #5, and according to the data sheet, it can be between 2.7 to 10 V. Pin #16 is the ground terminal.

The interconnection cross-reference scheme shown below is used as a checklist to each component's correct routing point.

C1	0.1 μF	from pin #1 to Vcc
C2	220 pF	from pin #18 to Vcc
C3	220 pF	from pin #17 to Vcc
C4	0.1 μF	from pin #15 to Vcc
C5	0.001 μF	from pin #14 to ground
C6	200 pF	from pin #13 to antenna
C7	100 pF	from pin #12 to Vcc
C8	0.001 μF	from pin #11 to Vcc
C9	100 pF	from pin #11 to pin #10
C10	0.01 μF	from pin #2 to ground
C11	0.1 μF	from pin #2 to audio amp
C12	0.01 μF	from pin #4 to Vcc
C13	0.01 μF	from pin #5 to ground
C14	220 pF	from pin #8 to Vcc
C15	0.001 μF	from pin #7 to #9
C16	10 to 20 pF	from pin #6 to Vcc
C17	0.01 μF	from D1/R2 to Vcc
C18	0.1 μF	from R2/R3 wiper to ground
R1	22 kohm	from pin #2 to ground
R2	100 kohm	from D1/C17 to C18/VR1 wiper
VR1	100 kohm	from Vcc to ground, wiper to R2/C18
D1	varicap diode	from C16/L1 to C17/R2
L1	tuning coil	from pin #6 to Vcc

Test Setup

The test setup is simplicity itself, since all the work is done by the IC. Couple the circuit output via capacitor C11 into a suitable audio power amplifier (such as the LM 386). Turn on the power to the amplifier first. Next apply power to the TDA 7000 receiver project and slowly adjust R3 to search across the FM broadcast band. If no mistakes have been made, you should be getting a pretty good signal. Once this has been established, you can begin optimizing the actual placement of the station coverage on VR1. The coil dimensions (turns and spacing tightness) can be adjusted, as together with C16 and D1 they form the resonant network that controls the station you're picking up. Remarkably, that's all there is to it. This prototype design performed reliably for years without any problems.

Parts List

Semiconductor
　　IC1: TDA 7000 Signetics FM receiver IC
　　IC2: LM78L05 regulator

Resistors (all resistors are 5 percent, 1/4 W)
- R1: 22 kohm
- R2: 100 kohm
- R3: 4.7 kohm

Capacitors
(All nonpolarized capacitors disc ceramic)
(All electrolytic capacitors have a 25 V rating)
- C1: 0.1 µF
- C2: 220 pF
- C3: 220 pF
- C4: 0.1 µF
- C5: 0.001 µF
- C6: 200 pF
- C7: 100 pF
- C8: 0.001 µF
- C9: 100 pF
- C10: 0.01 µF
- C11: 0.1 µF
- C12: 0.01 µF
- C13: 0.01 µF
- C14: 220 pF
- C15: 0.001 µF
- C16: 10 to 20 pF
- C17: 0.01 µF
- C18: 0.1 µF
- C19: 5 pF to 50 pF
- C20: 0.1 µF
- C21: 4700 µF

Additional materials
- VR1: 100 kohm potentiometer
- L1: Tuning coil
- D1: Varicap diode MV209 Motorola varicap tuning diode
- D2: LED
- S1: single-pole, single-throw miniature switch
- Power supply: nine-volt line adapter

CHAPTER **11**

FM Low-Power Transmitter: Project 40

Use a low-power FM transmitter to transmit audio signals from room to room.

Introduction

There's something fascinating about a transmitter circuit—the ability to communicate without the restrictions of interconnecting cables. You've no doubt seen some rock band guitar players leave the stage and move around and into the audience, untethered by conventional guitar cables. That roving ability is courtesy of a compact transmitter circuit that is strapped to the guitar. You can spot this transmitter by the absence of the usual long guitar cable. In its place is a short antenna protruding from the compact transmitter unit.

Professional transmitter units designed for tracking high-quality music signals often make use of a principle called "diversity reception," in which there is more than one source of the required signal being picked up. How is this done? This form of transmission/reception essentially revolves around sending the same (guitar) signal out with perhaps different frequencies, or sending out the same signal at different times. Whatever method is used, the purpose overall would be to compensate for the fading that takes place as a result of the fact that the transmitter's position is constantly changing.

Fading is just a definition of the weakening of the strength of received signal. Taken to the limit, when the signal becomes too weak it disappears altogether. Under normal circumstances, a broadcast transmitter is fixed (in terms of its location) at the base station, and more often than not, the receiver is also fixed (like the one in your home). But where the transmitter or receiver is moving (say, when you have the receiver in a moving automobile), you can experience fading in the received signal, especially where the transmitted signal is weak to begin with. For professional musicians, this fading cannot be tolerated, hence the utilization of diversity reception. For our FM low-power transmitter project described here, however, this is not an issue, as the target audience for this project is electronics enthusiasts rather than professional guitar players. Just bear in mind that signal fading, as the transmitter moves, is a natural consequence of the movement of either the transmitter or receiver relative to each other. You might get this with this setup—but it's a minor matter.

One type of popular transmitter circuit, the FM low-power transmitter (as described here), frequently finds its way in one guise or another into electronics hobby circuits. Amazingly, it requires the use of only one transistor, and a general-purpose one at that, to generate high-frequency signals in the VHF (88 to 108 MHz) band. This circuit enables any audio signal source to be easily frequency modulated, whereby the audio signal amplitude variations are translated into frequency variations. This is the final signal form that gets picked up by a standard broadcast receiver. This circuit example once more shows that an innovative circuit can be designed around a single active device (be it a transistor or an integrated circuit). You don't always need to have complex array upon array of devices to design circuits innovatively.

This low-power license-free transmitter is the cleanest example of such minimalist purity of design. It has everything you could want: simplicity of design, noncomplicated and easily obtainable components, impressive performance, and lots more! It's license free too, so there's no need to worry about the legality of the transmission. The type of circuit shown here is the same as that contained in popular kid's wireless microphones found in toy stores.

This project is a low-power FM transmitter circuit running off a nine-volt battery, sending out signals that can be picked up by any standard FM receiver that is tuned to the transmitter frequency. With just a short wire antenna, you can get a transmission capability from room to room and sometimes as far as between two adjacent houses. You could use this capability as the transmitting block for a baby monitor or to send your hi-fi audiotapes across your house without any trailing wires. The FM carrier transmitter is the heart of the system, and it starts by generating a carrier signal. The carrier signal (located in the VHF frequency region for this circuit) is the transmission medium by which any impressed audio information is transmitted, or "carried," across the airwaves. Once you've built the basic block, you can feed it with a variety of signal sources, such as an electric guitar signal or the output signal taken from the headphone socket of a cassette recorder.

Circuit Description

The heart of this FM low-power transmitter is just an ordinary, general-purpose NPN transistor, Q1 (2N3904), that can be readily found at any component store. The circuit is shown in Figure 11-1.

Although a 2N3904 is specified, that particular type of transistor is not critical; any other general-purpose NPN type can be used too. The circuit is configured as a high-frequency oscillator working in the frequency-modulated mode. That means that any audio signal feed causes the main carrier frequency to vary

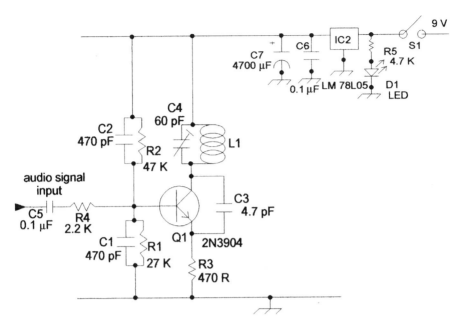

Figure 11-1 FM Transmitter

in step with the audio amplitude variations. A tuned circuit in the collector terminal resonates at a chosen (by the user) VHF frequency that is determined by the values set for the inductor and capacitor forming the tuned circuit. The tuned circuit is built up around a very simple home-brew coil (L1) and a small variable trimmer capacitor, C4 (60 pF). Neither the coil nor the capacitor is critical in terms of absolute values. Since the trimmer is variable and adjustable by the user, the transmitted carrier signal thus also varies.

The description of the circuit starts with a dc voltage bias set up that is provided by resistor, R1 (27 kohm), which is connected from the base terminal to ground and a second resistor, R2 (47 kohm), connected from the same base terminal to Vcc. With a nine-volt supply used, that translates to about a 3.3-volt bias applied to the base terminal. Resistors R1 and R2 form a straight potential divider, and the calculation for the voltage applied to the base terminal is given by:

$$V_{base} = 9 \text{ volts} \times [R1/(R1+R2)] = 9 \times [27/74] = 3.3 \text{ volts}$$

In parallel with each of the bias resistors are two capacitors, C1 (470 pF) and C2 (470 pF), which provide a bypass shunt for the high VHF frequencies. At a frequency of 100 MHz (which is our chosen reference midband frequency in the VHF FM band), the capacitive reactance is given by:

$$Xc = 1/(2 \times \pi \times f \times C) = 1/(2 \times 3.142 \times 100.10^6 \times 470.10^{-12})$$
$$= 1/(295.34.10^{-3}) = 3.38 \text{ ohms}$$

As you can see, this 3.38 ohms is essentially a short circuit. Since the circuit is designed to be an oscillator, there has to be a feedback path between the output and the input, and this is provided for by having a very small capacitor C3 (4.7 pF) shunting the collector and emitter terminals. The load current flowing through the transistor (via the collector) is limited by an emitter resistor, R3 (470 ohm). This is a fairly low-value resistance, optimized to sustain the oscillations and also to minimize the current drain from the supply voltage. The principal part of the circuit is the resonant inductor (L1) and capacitor trimmer (C4) combination in the collector path. The resonant frequency is given by: $f_{resonant} = 1/(2 \times \pi \times \sqrt{LC})$, where f is in units of hertz, L in henrys, and C in farads.

As the inductor is a home-brew affair, the inductance of L1 is a bit of an unknown. However, knowing the resonant frequency to be 100 MHz and the value of 2pF for C4 (from the table below), we can infer a value for L1 (should you wish to determine it). Rearranging the equation to calculate inductance, we get:

$$L = 1/(4 \times \pi^2 \times f_{resonant}^2 \times C) = 1/(4 \times 9.872 \times 10^{16} \times 2.10^{-12})$$
$$= 1/(78.976.10^4) = 1.266 \text{ microhenrys}$$

As a point of reference, this value of inductance is very small. The inductor or coil in the collector path forms a direct dc connection to the supply, hence there is a need for current limiting (provided for by the emitter resistor); otherwise, the current flowing through the collector terminal would be excessive. The antenna is just a short (six to twelve inch) section of solid hookup wire. Power is supplied through a nine-volt battery. The audio signal feed coming in via resistor R4 (2.2 kohm) and capacitor C5 (0.1 µF) causes a frequency-modulated signal to be produced, which is radiated via the antenna. This is the signal that is picked up by any standard FM broadcast receiver. Even without a modulating input signal, the carrier signal would be produced and radiated. When the FM receiver is tuned to the carrier frequency, the normally present hiss (no signal present) is silenced (signal present).

The stability of the frequency of oscillation is tied to the stability of the supply voltage—if the supply voltage changes, so too does the frequency. As the battery supply voltage drops, the resonant frequency falls. This is not a desirable characteristic, because the received signal will drift away from the setting on the FM receiver, that is, you'll lose the signal transmitted. A five-volt regulated supply is thus used to produce a stable supply. Where the input signal is large, though (e.g., if you were feeding in an electric guitar signal), there will be insufficient supply headroom (when using the reduced five-volt supply) for a high-amplitude signal input, and distortion will result. If you are using an electric guitar as a feed signal and wondering what sort of signal levels could be tolerated, here are some signal amplitude figures I've empirically determined. The source is a Fender Stratocaster feeding directly into resistor R4.

	Treble Strings (1,2,3)	Bass Strings (4,5,6)
Lightly struck	10 millivolts peak to peak	50 millivolts peak to peak
Heavily struck	40 millivolts peak to peak	200 millivolts peak to peak

Using a five-volt supply, the maximum peak-to-peak input voltage was found to be around 100 millivolts before distortion started to set in. That's the reason for having a higher supply voltage (e.g., nine volts); it gives you higher headroom for larger input-signal amplitudes. The supply current for this circuit is quite low, around five milliamps when using a nine-volt supply. This is a fairly reasonable supply voltage to use, as the nine-volt battery will give you quite a few hours of usage before the signal starts to drift. An alternate source for the signal input would be to use a cassette recorder as a source, taking the input signal feed from the cassette recorder's headphone output jack socket. Just adjust the cassette recorder signal output to give you the maximum strength signal with no distortion appearing at the receiver end. Maximizing the signal amplitude (before the onset of distortion) gives you the cleanest received signal quality, with the lowest noise contribution.

Stabilizing RF Oscillator Drift

The quality of any RF-based device, be it receiver, transmitter, or signal generator, can be defined by how stable the frequency of oscillation is. Here are three general tips that will improve the operation of any high-frequency oscillator build. RF oscillators operating in the higher frequency region (tens of megahertz and upwards) will drift in frequency as a result of some worst-case construction/layout conditions. At a low oscillator frequency, this drift in frequency might be tolerable, but for reliable long-term high (oscillator) frequency operation, drift is a no-no! The popular, license-free FM transmitter design (this project) uses a single transistor operating essentially around 100 MHz. A 1 percent drift in oscillator frequency, for example, translates to a 1 MHz shift in frequency, which is quite considerable. There are several "good" construction/layout techniques listed below that minimize and control the unwanted frequency drift.

1. One of the most critical influences on frequency stability is temperature, particularly high temperatures and temperature variations. Therefore, locate the oscillator away from any component that is generating excessive heat, such as regulators and power transistors that consume high current.
2. Use an IC regulator that supplies only the oscillator circuit, so that its supply line is not affected by any of the other circuit components. A circuit block, such as a power amplifier feeding into a low-impedance load, will cause the supply current and hence the supply voltage to fluctuate as the load current changes. This change, if

allowed to "interfere" with the oscillator's power supply, will lead to an undesirable drift in oscillator frequency.
3. Keep the frequency-controlling components firmly mounted on the circuit board. In the simple FM transmitter design, the frequency-associated components are a capacitor and an inductor. The capacitor is generally a small trimmer and is not much of a problem. The inductor, however, which is often a small, hand-wound coil, will cause a change in frequency if the turns are allowed to move relative to each other. Make sure therefore that the construction is firm—use epoxy if required to keep the turns in place, and make sure the coil cannot move, especially if the project is a portable device.
4. Insulate the RF oscillator from ambient temperature variations, with a "box" made from polystyrene sheet. Polystyrene is an excellent heat insulator. The sheets can be cut quite easily to shape with a hobby knife and glued together with contact cement. The heat-shielding properties of a "poly" cover are quite considerable.

Construction Tips

There are relatively few components used in this project, so a small section of assembly board is all that is needed. The transistor, Q1, is mounted first, taking care that the correct emitter, base, and collector leads are identified correctly. To get the transistor positioned correctly, if you're normally used to handling ICs only, start with the assembly board oriented in such a way that the power rail is at the top and the ground rail at the bottom. The transistor should be positioned with the following convention: emitter facing south, base facing west, and collector facing north. This convention makes it easier to later check on your component placement accuracy. The transistor leads are fairly robust, though short. Use a pair of small needle-nosed pliers to bend the leads as required. The particular type of assembly board you choose will, of course, determine the actual arrangements for Q1. Regardless, the board tracks need to be isolated for the three transistor leads. Once Q1 is soldered in place, start the construction with the placement of the two resistors (R1, R2) used for the base terminal and of the resistor (R3) for the emitter terminal. There will be plenty of space on the board. Next add the capacitors (C1, C2) across the base resistors, and then the capacitor (C3) bridging the collector and emitter terminals. The input feed resistor (R4) and capacitor (C5) can be added next. You'll see that what you have laid out on the assembly board closely resembles the circuit schematic itself. That way, checking for errors arising from misdirected connections is much easier. A dedicated printed circuit board, on the other hand, makes circuit tracing an arduous task, as you need to trace each individual connection from the underside—and the traces are not nice and linear. That's one advantage for using prototyping assembly boards—you've got a linear grid of solder tracks to work with.

FM Low-Power Transmitter: Project 40

The inductor (L1) and trimmer capacitor (C4) are added next. Note that the trimmer capacitor (C4) will be subjected to much usage as you search through the frequency band. That means its mechanical mounting arrangements should be very securely made to the assembly board. The rotor (movable) section of the trimmer capacitor can be fairly stiff, and you don't want the soldered leads to take up the stress from numerous mechanical rotations. A short wire antenna is soldered directly to the collector terminal. There will be flexing of this wire antenna, so anchor it to the board at a point a short distance away from the solder connection. It's a good idea also to minimize the heat applied to the transistor (excess heat can damage the device) when making the solder connections, by effecting the solder joint as quick as possible. This is particularly important at terminals where there are many components connected—as in the case of the base terminal.

The coil, L1, is very easily constructed. Start with a length of regular solid hookup wire, start with one end and wrap it around a pencil. The coil dimensions are about six turns (not critical) of solid hookup wire (with the insulation left intact) wound with a diameter equal to the pencil we're using, with the turns well spaced (about a wire diameter apart), and with the total turns when you're finished about one inch in length. Cut the coil free and strip of the ends of the insulation. Compared to the rest of the components, L1 is fairly large, so allow adequate mounting space for it.

The five-volt regulator (IC1) is part of the power-supply section and fed with a nine-volt battery as its power source. The power on/off switch (S1) will be mounted off board, and on a project case, together with power indicator LED (D1) and LED current limiter resistor (R5). Capacitor C6 is on the input side to IC1. Capacitors C7 and C8 are on the output side of IC1. Take care that the polarity of C8 is correctly mounted.

Test Setup

You'll need a standard FM receiver for the test setup. Tune the receiver to a quiet spot somewhere on the dial, somewhere around 100 MHz. The actual frequency is not critical, as all we're looking for is a spot somewhere in the middle of the band and where there is no broadcast station. There should be just a hissing sound coming from the speaker—this is the noise you hear in the absence of a station. Turn on the transmitter circuit, keeping the receiver a few feet distance away. No input signal to the transmitter is required at this stage of the test, because we're first going to tune the resonant frequency of the transmitter to match the receiver frequency (our "quiet spot").

A trimmer tool is used to adjust the trimmer capacitor. This trimmer tool looks like a thin pencil with a thin, recessed metal blade at the tip that's designed to turn the slot on the top of the trimmer. At a pinch you can use a small metal screwdriver, but being metal, there's going to be a hand-capacitance effect. That is, when you remove the screwdriver, the frequency will change, because your hand capacitance will have influenced the tuning. It's better to use the proper tool, as you're going to be searching very carefully for the fundamental frequency and not the weaker second harmonic frequency.

Carefully and very slowly rotate the trimmer with the trimmer tool. The aim of this search is to null out the receiver hiss. This occurs when the transmitter is tuned (by adjusting the trimmer) to the same frequency as the receiver is tuned to—the circuit will now be transmitting a carrier frequency at that receiver frequency. Once you've located a carrier signal, take note of the position of the trimmer's adjusting slot, because we want to return to this location later, and go carefully through the entire trimmer rotation to verify the presence of the two fundamental and second harmonic frequencies. Once you've found these, take note of where they're located. The positions should be similar to that shown below, in that there'll be two locations for the stronger fundamental and two locations for the weaker second harmonic. The sequence will be fundamental followed by second harmonic and then second harmonic followed by fundamental, as the rotor is adjusted in a clockwise direction. As the trimmer is a mechanical device, there will also be mechanical backlash present, so if you feel you've gone past a resonant point, continue turning the rotor in a clockwise direction to return to that spot, rather than going backwards to find it.

Most of the initial problems I encountered (and it took quite awhile) were with trying to find the resonant frequency (with the trimmer), not knowing if I was in the correct area (of the frequency spectrum). There were times when I wondered whether the circuit was functioning, because evidence of the carrier is the only way of knowing. Going too rapidly through the trimmer's span will produce essentially no signal, because the mechanical position at which the correct frequency is found is extremely narrow. It is very easy to go past the resonant frequency without knowing it. Using a small mechanical variable capacitor trimmer is almost a hit-and-miss affair, as you don't know what the capacitance versus the vane position is.

As an aid to what is a tedious mechanical search for the resonant point, I've empirically generated some test results for carrier frequencies versus trimmer's rotor positions. The resonant frequency was measured by retuning the receiver each time to the new generated carrier and noting the indicated frequency. I've found this table to be extremely helpful in finding your way around the trimmer's position versus frequency. It's also useful as a reference for locating other resonant frequency points. Notice that the same value of capacitance can be found at two positions. This is a result of the mechanical construction of the overlapping trimmer vanes (one stationary and one rotating), causing any particular overlap to occur at two positions. This means that our reference frequency of 100 MHz can be found at two positions. There is also the 50 MHz carrier, which will produce a weaker second harmonic signal at our desired 100 MHz spot. There is a large rate of change of frequency with a very small rotation of the vane. All these considerations mean it is very easy to miss the expected resonant frequency. Hence go very slowly as you rotate the trimmer.

The table below gives you some gross (but still sufficiently informative) indications of the empirical values of inferred capacitance associated with the trimmer's vane position. The flat position refers to the rotor profile on the particular type of trimmer used, and it is used as a convenient point of reference. The flat rotation direction is clockwise when these measurements were taken.

Flat position	Capacitance	Resonant Frequency
9 o'clock	1 pF	—
10 o'clock	2 pF	100 MHz (fundamental)
11 o'clock	4 pF	50 MHz (2nd harmonic)
1 o'clock	6 pF	—
2 o'clock	8 pF	—
4 o'clock	8 pF	—
5 o'clock	6 pF	—
6 o'clock	4 pF	50 MHz (2nd harmonic)
7 o'clock	2 pF	100 MHz (fundamental)
8 o'clock	1 pF	—

Parts List

Semiconductors
    Q1: NPN 2N3904 general-purpose transistor
    IC2: LM78L05 regulator
Resistors (all resistors are 5 percent, 1/4 W)
    R1: 27 kohm
    R2: 47 kohm
    R3: 470 ohm
    R4: 2.2 kohm
    R5: 4.7 kohm
Capacitors
(All nonpolarized capacitors disc ceramic)
(All electrolytic capacitors have a 25 V rating)
    C1: 470 pF
    C2: 470 pF
    C3: 4.7 pF
    C4: 60 pF trimmer capacitor
    C5: 0.1 µF
    C6: 0.1 µF
    C7: 4700 µF

Additional materials
   L1: home-brew coil
   D1: LED
   S1: single-pole, single-throw miniature switch
   Power supply: nine-volt battery

Index

Numerals
2N2222 transistor, 134
2N3055 transistor, 137–139
5400 series, 26
7400 series. *See* TTL (transistor-transistor logic)
74LS14 Schmitt trigger inverter
   characteristics of, 97
   debouncer circuit using, 108–110
   pulse oscillator using, 97–100
   triangle wave generator using, 105–108
   turn-off delay circuit using, 103–105
   turn-on delay circuit using, 100–102
74LS75 quad latch demo
   circuit description, 126–128
   construction tips, 128
   function of, 126
   parts list, 128–129
   testing, 128
74LS122 monostable demo
   circuit description, 123–125
   construction tips, 125
   parts list, 125–126
   pulse width calculation, 123–124
   testing, 125
14000 series, 20, 22
14001 NOR gate
   characteristics of, 111
   demo of, 111–114
   latch using, 117–120
   metronome using, 120–123

A
AC-DC conversion, 13–14
accelerometers, 10
amplifiers, 12
analog electronics, 9–11
analog meters, 32, 34
analog signals, 7, 9, 10–11
analog testing, 27
AND gates, 23–24, 45–48
application note circuits, 123

B
base, 18
BCD coding (binary coded decimal), 17–18
binary system, 17, 18–19
bistables, 126
BNC sockets, 30

C
capacitors
   axial-lead, 67
   basics of, 66
   effect on digital signal, 8
   marking codes on, 66–67
   nonpolarized, 67
   parallel-connected, 67–68
   polarized, 67
   radial-lead, 67
   series-connected, 67–68
   trimmer, 68
   units rated in, 66

capacitors—*continued*
  variable, 68
  working voltage rating, 68
CD4017 decade counter
  features of, 81
  pulse speed reducer using, 81–84
CD4072 quad input OR gate,
    129–131
CMOS (complementary metal-oxide
    semiconductor) devices, 19–22,
    83, 111
color code, resistor, 63
comparator, 86
complementary metal-oxide
    semiconductor (CMOS)
    devices, 19–22, 83, 111
counter. *See* CD4017 decade counter
crystal sets, 8, 181
current flow, 33
current measurements, 33–34

D
dc adapters, 153–154
  See also line adapters
dc level shifter
  circuit description, 93–94
  construction tips, 94–95
  parts list, 96
  purpose, 93
  testing, 95
dc wall transformers. *See* line adapters
debouncer circuit. *See* Schmitt switch
    debouncer
decade counter. *See* CD4017 decade
    counter
decimal system, 18–19
decoupling, 26
digital signals, 11
diodes
  protective, 136
  Zener, 173
division, 36
driver circuits
  buffer for, 143–146
  high-power FET, 141–143
  high-power transistor, 137–140
  low-power TTL relay, 133–137
duty cycle, 42

E
electrostatic damage, 21
EX NOR gates, 56–57
EX OR gates, 55–56

F
fading, 189–190
farads, 66
FET (field effect transistor), 141–142
field effect transistor (FET), 141–142
filtering, 15–16
fluorescence, 35
FM receiver chip, 183–184
frequency measurements, 37–38
full-wave rectification, 13
function generators
  amplitude control, 40
  basics of, 39–40
  dc offset control, 40
  duty cycle control, 40
  frequency range control, 40
  sweep frequency control, 41
  TTL/CMOS switch, 40–41

G
gain, 15, 31
generating signals, 14–15

H
half-wave rectification, 13
harmonics, 14
hi-lo trip detector
  circuit description, 88–89
  construction tips, 89
  parts list, 90
  purpose, 88
  testing, 90
high-cut filters, 15–16
high-low state display
  circuit description, 78–79
  construction tips, 80
  output frequency calculation, 79
  parts list, 80–81
  purpose, 77
  testing, 80

I
IFR 610 field effect transistor, 141–142

integrated circuits
  basic analog, 22
  current draw of, 30
  most useful, 14
  stability of, 12

J

jack plugs, 70, 168
jack socket-power switch
  circuit description, 171–172
  construction tips, 172
  parts list, 173
  purpose of, 171
  testing, 173
jack sockets, 70, 166–168

L

latching circuits
  defined, 115
  using 74LS75 quad latch, 126–129
  using NOR gate, 117–120
  using SCR, 114–117
line adapters, 13–14, 157
  See also dc adapters
LM 324 quad operational amplifier
  dc mode, 85
  driver buffer using, 143–146
  features of, 85–86
  high-lo trip detector using, 88–90
  threshold detector using, 84–88
  voltage indicator using, 148–152
LM 386, 9, 14, 22
LM 555 timer
  astable mode, 72–73
  features of, 70–71
  high-low state display using, 77–81
  modes of, 14
  oscillator using, 41–44
  output frequency calculation, 75–76
  pulse generator using, 71–77
LM 741 operational amplifier, 93–94
LM 78L05 voltage regulator, 73
LM 78L12 voltage regulator, 154
LM 78L62 voltage regulator, 155
loads, 13
logic states, 17
low-pass filters, 15–16

low-power Schottky TTL (LSTTL), 20–22

M

MC 14000 series, 20, 22
MC 14001. *See* 14001 NOR gate
meter overload protector
  circuit description, 169–170
  construction tips, 170
  need for, 169
  parts list, 171
  testing, 170–171
metronome
  circuit description, 120–122
  construction tips, 122
  parts list, 122–123
  testing, 122
microfarad, 66
microphones, 9–10
monostables, 123–125
multimeters, 31–34

N

NAND gates, 50–53
nanofarad, 66
noise, 26
NOR gates
  demo of
    circuit description, 112–113
    construction tips, 113
    parts list, 114
    testing, 114
  latch using
    circuit description, 117–118
    construction tips, 119
    parts list, 119–120
    testing, 119
  truth tables for, 53–55

O

"off" period, 42
ohm, 60
Ohm's Law, 61
"on" period, 42
operational amplifiers, 31, 85
OR gates
  switcher using, 129–132
  truth tables for, 48–50

202 BEGINNING DIGITAL ELECTRONICS THROUGH PROJECTS

oscillators
   characteristics of, 7–8
   LM 555 demo, 41–44
   Schmitt pulse, 97–100
oscilloscopes
   basics of, 35–36
   brightness control, 38
   cabling to/from, 30
   calibration signal source, 38
   coupling signals to, 38
   external trigger control, 39
   focus control, 38
   gain controls, 36–37
   shift controls, 38
   slope selector, 39
   stabilizing display, 39
   sweep rate control, 37
   timebase, 37
   trigger control, 38–39
   waveform value measurements with, 43–44

P

phone jacks, 70, 166–168
phone plugs, 70, 168
phosphor screen, 35
photoconductive cells, 10
picofarad, 66
plug transformers. *See* line adapters
plugs, 70, 168
polarity protector
   circuit description, 161–162
   construction tips, 162
   need for, 161
   parts list, 162–163
   testing, 162
poles, 69
potential divider circuit, 61
potentiometers, 61–62
power, 13
power amplification, 13–14
power supplies
   5-volt
      circuit description, 158
      construction tips, 158
      need for, 157
      parts list, 159
      testing, 158
   6.2-volt ultra stable
      circuit description, 154–155
      construction tips, 155
      need for, 153–154
      parts list, 156–157
      testing, 155–156
   basics of, 60
   battery or adapter, 159–161
   capacitors in, 60
   decoupling, 26
   function of, 13
   resistors in, 60
power switch, socket-activated, 171–173
pre-amplification, 12–13
pre-amplifiers, testing, 29–31
pulse generator
   circuit description, 72–75
   construction tips, 76
   features of, 71–72
   parts list, 77
   testing, 76–77
pulse oscillator. *See* Schmitt pulse oscillator
pulse speed reducer
   circuit description, 81–83
   construction tips, 83
   parts list, 84
   purpose, 81
   testing, 83
pulse stretcher
   circuit description, 91–92
   parts list, 92–93
   purpose, 90–91
   testing, 92

R

radix, 18
RC integrator/differentiator, 174–177
reactance, 16
receivers
   FM band
      circuit description, 183–184
      construction tips, 184–187
      overview of, 183
      parts list, 187–188
      testing, 187
   primitive, 181
   superheterodyne, 182–183
   tuned radio frequency, 182
rectification, 13

regulator protector
    circuit description, 163–164
    function of, 163
    parts list, 164
relay driver circuit
    circuit description, 134–136
    construction tips, 136
    parts list, 137
    purpose, 133–134
    testing, 136–137
relays, 135–136
resistance measurements, 34, 62–63
resistors
    color code for, 63
    defined, 60
    function of, 60–61
    parallel-connected, 64
    potential-dividing, 61
    power ratings of, 65
    in power supplies, 60
    series-connected, 63–64
    tolerances of, 64–65
    variable, 61–62
RF oscillator 6.2-volt power supply
    circuit description, 154–155
    construction tips, 155
    need for, 153–154
    parts list, 156–157
    testing, 155–156
ripple, 13

S
Schmitt pulse oscillator
    circuit description, 98–99
    construction tips, 99
    output frequency calculation, 97
    parts list, 100
    purpose, 97–98
    testing, 99
Schmitt switch debouncer
    circuit description, 109–110
    construction tips, 110
    parts list, 110
    purpose, 108
    testing, 110
SCR (silicon-controlled rectifier) demo
    circuit description, 115–116
    construction tips, 116
    function of, 114
    parts list, 116–117
    SCR characteristics, 115, 116
    testing, 116
signals
    analog, 7, 9, 10–11
    digital, 11
    filtering, 15–16
    generation of, 14–15
    injection of, 15
    power amplification of, 13–14
    pre-amplification of, 12–13
    sources of, 12
socket power switch, 171–173
sockets
    IC, 23
    jack, 70, 166–168
square waves, 10–11, 14
static electricity damage, 21
switcher demo
    circuit description, 130–131
    construction tips, 131
    parts list, 131–132
    purpose, 129
    testing, 131
switches, 69

T
TDA 7000 FM receiver chip, 183–184
test equipment
    basic suite, 28
    function generators, 39–41
    multimeters, 31–34
    oscilloscopes, 35–39
testing
    general procedures, 29–30
    pre-amplifier, 30–31
thermistors, 10
threshold level detector
    See also TTL driver buffer
    circuit description, 85–87
    construction tips, 87
    parts list, 87–88
    purpose, 84
    testing, 87
throw, 69
time-delay circuits
    turn-off Schmitt, 103–105
    turn-on Schmitt, 100–102
transducers, 9–10

transistor drivers, high-power
  2N3055
    circuit description, 138–139
    construction tips, 139–140
    function of, 137–138
    parts list, 140
    testing, 140
  IFR 610 FET
    circuit description, 141–142
    construction tips, 142
    parts list, 143
    testing, 142–143
transistor-transistor logic. *See* TTL
transistors, 134–135
transmitter
  circuit description, 190–193
  construction tips, 194–195
  parts list, 197–198
  testing, 195–197
  uses of, 190
triangle wave generator
  circuit description, 105–107
  construction tips, 107
  frequency calculation, 106
  parts list, 108
  purpose, 105
  testing, 107
triangular waves, 14
trimmer capacitors, 68
troubleshooting, 28–31
truth tables
  AND gate, 45–48
  defined, 45–46
  EX NOR gate, 56–57
  EX OR gate, 55–56
  NAND gate, 50–53
  NOR gate, 53–55
  OR gate, 48–50
TTL (transistor-transistor logic)
  advanced low-power Schottky family (ALS), 25
  advanced Schottky family (AS), 25
  characteristics of, 19–21, 24
  family development, 24–25
  frequently used devices, 25–26
  low-power family (L), 24
  low-power Schottky family (LS), 24
  Schottky family (S), 25
  supply voltage, 19, 20
  temperature range, 26
TTL driver buffer
  circuit description, 144–145
  construction tips, 145
  parts list, 146
  purpose, 143
  testing, 145–146
tubes, 8–9
turn-off delay Schmitt
  circuit description, 103–104
  construction tips, 104
  function of, 103
  parts list, 104–105
  testing, 104
turn-on delay Schmitt
  circuit description, 100–101
  construction tips, 101–102
  function of, 100
  parts list, 102
  testing, 102

U
unused inputs, 20

V
vacuum tubes, 8–9
variable resistor substitute, 177–179
voltage divider circuit, 61
voltage level indicators
  active
    advantage of, 148–149
    circuit description, 149–151
    construction tips, 151
    parts list, 152
    testing, 152
  passive
    circuit description, 146–147
    construction tips, 148
    function of, 146
    parts list, 148
    testing, 148
voltage measurements, 32–33
voltage range indicator. *See* hi-lo trip detector
voltage regulator circuits
  using 78L05, 73, 74
  using Zener diode, 173–174
voltage regulator ICs

LM 78L05, 73
LM 78L12, 154
LM 78L62, 155
   protective diodes for, 163–164
voltage source, reversible, 164–166
volume, 13
V_{IH}, 22
V_{IL}, 22

W
waveform differentiator, 174–177
waveform integrator, 174–177
waveform value measurements, 43–44

Z
Zener diode voltage regulator, 173–174
Zener diodes, 173